Optical Polymers

ACS SYMPOSIUM SERIES **795**

Optical Polymers

Fibers and Waveguides

Julie P. Harmon, Editor
University of South Florida

Gerry K. Noren, Editor
DSM Desotech, Inc.

American Chemical Society, Washington, DC

Library of Congress Cataloging-in-Publication Data

Optical polymers : fibers and waveguides / Julie P. Harmon, Gerry K. Noren,, editors.

   p. cm.—(ACS symposium series ; 795)

From a symposium presented at the 218th ACS National Meeting in New Orleans Aug. 1999.

Includes bibliographical references and index.

ISBN 0–8412–3706–9

   1. Optical wave guides—Materials—Congresses. 2. Optical fibers—Materials—Congresses. 3. Polymers—Optical properties—Congresses. 4. Plastic lenses—Congresses.

   I. Harmon, Julie P., 1949- II. Noren, Gerry K., 1949- III. Series.

TA1750 .O557   2001
621.36′92—dc21   2001022570

Foreword

The ACS Symposium Series was first published in 1974 to provide a mechanism for publishing symposia quickly in book form. The purpose of the series is to publish timely, comprehensive books developed from ACS sponsored symposia based on current scientific research. Occasionally, books are developed from symposia sponsored by other organizations when the topic is of keen interest to the chemistry audience.

Before agreeing to publish a book, the proposed table of contents is reviewed for appropriate and comprehensive coverage and for interest to the audience. Some papers may be excluded to better focus the book; others may be added to provide comprehensiveness. When appropriate, overview or introductory chapters are added. Drafts of chapters are peer-reviewed prior to final acceptance or rejection, and manuscripts are prepared in camera-ready format.

As a rule, only original research papers and original review papers are included in the volumes. Verbatim reproductions of previously published papers are not accepted.

ACS Books Department

Contents

Preface

Light harvesting is an integral part of many scientific disciplines and is of paramount importance in many economically important applications. Recent research has focused on the use of polymers in optical fibers and waveguides. Polymers have advantages over conventional glass materials in that they are low cost, lightweight, flexible, and can be processed into large diameter fibers. Great advances have been made in the development of novel local area network systems, high-speed fibers, light guides, and displays. Polymers, unlike silica materials, offer the advantage of being easily doped with organic molecules. Dye-doped polymers have been used to construct scintillators used to detect high-energy radiation and in other fluorescent devices used to produce sensors, switches, and modulators. All of these achievements have presented us with enormous challenges, because they have evidenced the need to develop devices with stable refractive index gradients, increased transparency throughout the UV, visible and near infrared region of the electromagnetic spectrum and increased temperature resistance.

This volume was developed from a symposium presented at the 218th American Chemical Society (ACS) National Meeting, titled: "Optical Polymers: Advances in Optical Fibers and Waveguides" in New Orleans, Louisiana, August 1999. The ACS Division of Polymer Chemistry, Inc. sponsored this symposium. The purpose of the symposium was to bring together scientists from different disciplines to discuss the future needs of polymers in optical applications.

This volume provides an overview of recent research in several areas of fiber–waveguide technology. Among the main topics covered are a tutorial on optical fibers and a discussion on the feasibility of overcoming drawbacks associated with plastic optical fibers systems. Additional chapters describe novel polymers for nonlinear optical applications; optically active, erbium doped polymers; novel, fluorinated polymers, blends, and UV-curable polymeric cladding materials; hard plastic cladding materials; fluorescent optical fibers for data transmission and scintillator polymers for detectors; design of gradient index fibers; and the effect of climatic and mechanical environments on transmission in fibers.

This book is interdisciplinary in nature and is useful to physicists, chemists, engineers, materials scientists, as well as individuals and organizations concerned with synthesis, design and processing of waveguides.

Acknowledgments

Contributors to the symposium and to this volume are thanked for sharing their work and scientific expertise with all who are interested in polymeric waveguides. In addition, the editors acknowledge the ACS Division of Polymer Chemistry, Inc. and Kelly Dennis at ACS Books Department for her help with this volume.

Julie P. Harmon
Department of Chemistry
University of South Florida
SCA 400
4202 East Fowler Avenue
Tampa, FL 33620–5250

Gerry K. Noren
Fiber Optic Materials Research
DSM Desotech Inc.
1122 Saint Charles Street
Elgin, Illinois 60120

Chapter 1

Polymers for Optical Fibers and Waveguides: An Overview

Julie P. Harmon

Chemistry Department, University of South Florida, 4202 East Fowler Avenue, Tampa, FL 33620–5250 (harmon@chuma1.cas.usf.edu)

A brief discussion of the basics of fiber optics and methods of plastic fiber production is presented. Light attenuation is discussed with reference to specific optical polymers and extrinsic and intrinsic factors, which effect loss in fibers and waveguides. Structure -property relations are used to identify polymers with controlled refractive indexes, transparency in the Near-IR region of the electromagnetic spectrum and increased resistance to high temperatures. This overview also includes information on polymeric cladding and coating systems for glass core materials. The chapter ends with a discussion of important applications for plastic fibers and waveguides.

Plastic optical fibers (POFs) and waveguides have been of interest since the late 1960s and early 1970s (1). Plastics offer the advantages of being lightweight, flexible and easy to handle. High flexibility allows fibers of 1mm diameter to be produced and easily connected. The variable chemistry of polymers also makes it possible to design fibers and waveguides with high light harvesting capabilities. These advantages are, however, offset by intrinsic and extrinsic loss mechanisms causing high attenuation as compared to conventional glass fibers. The contributions to this book and other symposia and reviews trace the history of plastic optical fibers from early to *state of the art* work. This symposium series is aimed at bringing fiber and waveguide designers and fabricators in contact with polymer scientists. There is a plethora of recent, intricate fiber designs and applications. If both groups of scientists merge their technology there is almost a limitless range of optical waveguides on the horizon.

Design of Fiber Waveguides

The basics of fiber optics and waveguides have been discussed in detail (2-5) and specifically for POFs (6). Most of this chapter focuses on optical fibers. However, optical fibers are a specific type of optical waveguides; there are two other fundamental types of waveguides (3). The simplest waveguide is a planar waveguide consisting of a film of thickness comparable to the wavelength of light being propagated. The film material has a refractive index greater than that of the substrate on which it is deposited. A lower refractive index coating or cladding material is deposited atop the film. Another waveguide type is the channel waveguide whose cross-section in the x-y plane is rectangular with the axis of propagation in the z direction. X-y dimensions are of the order of microns. Propagation distances may extend over many cms. Again, the surrounding media have lower refractive indexes that the channel core. The following discussion pertains to optical fibers.

Snell's Law, Refraction and Reflection:

Light travels along a fiber by total internal reflection. This phenomenon arises from optimization of the difference in refractive indexes between the fiber core material and a thin layer of cladding material, which coats the fiber. The index of refraction of a material is the ratio of the velocity of light in vacuum to the velocity of light in the material, and is abbreviated, n. When a ray of light passes from the core material with a refractive index, n_1, to the cladding material with a lower refractive index, n_2, the light is refracted away from the normal. The angle of incidence is increased until the refracted ray is 90° with the normal. This angle is termed the critical angle. All light that enters the fiber at an angle greater than the critical angle, θ_{cr}, is reflected through the fiber core. θ_{cr} is deduced from Snell's law:

$$\theta_{cr} = \sin^{-1} (n_2/n_1)$$

Light is confined in the core if it strikes the cladding interface at an angle of (90° - θ_{cr}) or less to the surface. The maximum value of (90° - θ_{cr}) is the confinement angle, θ_{co}. Light entering the fiber from the outside is characterized by an acceptance angle, which differs from the confinement angle, since the light is entering from air. The acceptance angle, θ, is defined (5):

$$\theta = \text{Sin-1 NA}$$

Where NA is the numerical aperture defined as:

$$NA = \{[\, n_1^2 - n_2^2]^{1/2}\}/ n_0$$

Where n_0 is the refractive index of the intermediate medium, 1.0 for air.

As light travels down a circular fiber, the acceptance angle, θ, defines an acceptance cone, shown in fig. 1. As NA increases, the acceptance cone increases; the fiber transports light from a broader field.

This discussion applies to step-index fibers, fibers composed of a core with a step change in refractive index at the cladding interface. There are, however, different types of optical fibers based on the core-cladding refractive index profile and fiber dimensions.

Fiber Construction:

Fibers are classified according to their composition or by the way in which light propagates through the fiber. There are three types of fibers based on materials composition: glass core/glass clad fibers, glass core/polymer clad fibers and polymer core/polymer clad fibers. Although most of this overview is concerned with polymeric fibers, both types of glass fibers make use of polymers as protective coatings. Polymeric protective coatings are applied at the time of fiber drawing and are designed to maintain fiber strength (7). Typically, a soft primary coating is followed with a harder secondary coating. The overall thickness of the coatings is from 62 to 187 microns (8). Fiber manufacturers have devoted much research to the optimization of these coatings, since when fibers are exposed to temperature fluctuations, interfacial stresses are induced on the fiber due to the mismatch in coefficients of thermal expansion between the glass and coating polymers. Glasses have expansion coefficients on the order of $10^{-7}/°$ as compared to $10^{-4}/°$ for plastic (9). While this mismatch may improve coating adhesion, thermal stresses often result in light propagation losses. Methods designed to minimize stresses involve the following manipulations (7, 10):

1. Increasing the Poisson's ratio of the primary coating
2. Increasing the Young's modulus of the secondary coating
3. Decreasing the Young's modulus of the primary coating
4. Decreasing the thickness of the secondary coating

Soft polymers with T_gs below 0 °C are typically used for primary coatings; examples are polyethers, polyurethanes and polyacrylates. Secondary coatings typically have glass transition temperatures greater that 50 °C examples are polyarylates, polyamides, polytetrafluoroethylene and polyvinyl chloride (9).

Figure 2 illustrates fiber classification based on light propagation mechanisms. (For a more in depth discussion of this topic, the reader is referred to references 2 and 5.) Basically, the refractive index profile in the fiber core and the mode of propagation determine the efficiency of light propagation in the fiber. A mode is the path that light takes when it travels through a fiber. The different modes are depicted in figure 2. The refractive index profile determines the possible modes that the light rays assume. Step-index fibers may be single or multi-mode fibers. In multi-mode, step-index fibers, light may travel in different modes; several rays emanating from the same point of origin travel down the fiber and reach the opposite end of the fiber at different times. When the input light is in the form of a pulse, the output pulse is spread and this is termed modal dispersion. Single-mode fibers are constructed with such narrow cores that only one mode of light travel is possible; typical core sizes may be less that 10 microns. This is one method of reducing modal dispersion. Another method of reducing pulse dispersion involves using fiber cores with a gradient in refractive index that slowly increases from the cladding to the fiber core.

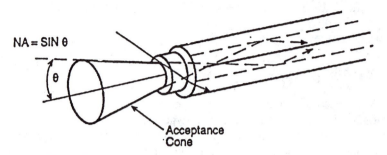

NA = SIN θ

θ

Acceptance Cone

Figure 1. Acceptance cone in an optical fiber, reference 5. Reprinted with the permission of Tyco Electronics Corporation.

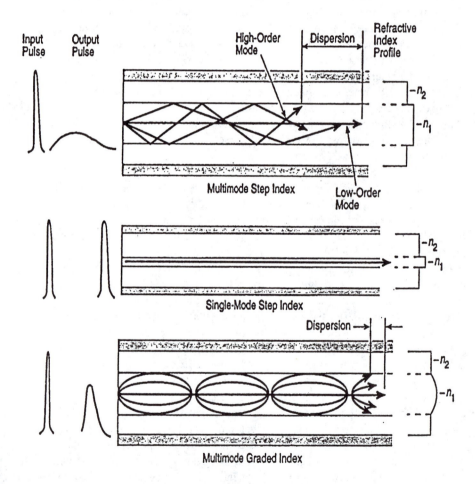

Figure 2.Light paths and outputs in different fiber types, reference 5. Reprinted with the permission of Tyco Electronics Corporation.

The speed of light is faster in the lower index material and individual rays of light reach the final destination with minimum dispersion.

Fiber designers focus on reducing light loss in fibers in two ways, limiting dispersion and limiting attenuation. Loss is measured in units of the decibel, dB, which is related to the input power, P_i and output power, P_o (4):

$$dB = -10 \log (P_o/P_i)$$

The above discussion of fiber design focussed on pulse broadening due to intermodal dispersion. Pulse broadening in POFs is also the result of chromatic dispersion (11). Chromatic dispersion results from the wavelength dependency of the refractive index of the core material; different wavelengths of light travel at different speeds down the fiber core. The ultimate goal is to search for materials that have refractive indexes that are independent of wavelength. In multimode fibers the dispersion is quantified as a bandwidth length product, BWL. BWL is the signal frequency that can be transmitted through a specified distance and is expressed in terms of kHz-km. BWL is commonly specified as the point at which the signal strength drops by 3 dB (2). Single-mode fibers quantify dispersion at a particular wavelength in picoseconds per kilometer per nanometer (5).

Attenuation losses are discussed later in this overview.

Fiber Processing

Step-Index Fibers:

Step-index fibers are processed by coextrusion or by drawing preforms (12-14). The preform method was adopted from methods used to pull conventional glass fiber (15). Both processing methods require the use of highly purified polymer, usually produced by vinyl polymerization. The preferred method of synthesis for these materials is bulk polymerization. Emulsion and suspension polymerization techniques employ the use of additives that remain in the polymer or on the surface of particles and contribute to power losses in the fibers. Monomers are purified by distillation in a closed system. Chain transfer agents are added to control molecular weight; high molecular weight polymers undergo melt fracture during processing as a result of the formation of entanglements. Initiator may be added to the monomer prior to polymerization, while monomers that spontaneously undergo thermal polymerization are preferred.

The fiber drawing and extrusion processes are illustrated in figure 3. The preform is composed of core and cladding material with dimensions 30-40 times the size of the final fiber (6). Preforms are constructed by polymerizing the core material and coating the preform core with a solution of cladding prepolymer that is then fully polymerized. An alternate method is to polymerize a cylinder of the core polymer and a tube of the cladding polymer separately and to assemble the preform after

6

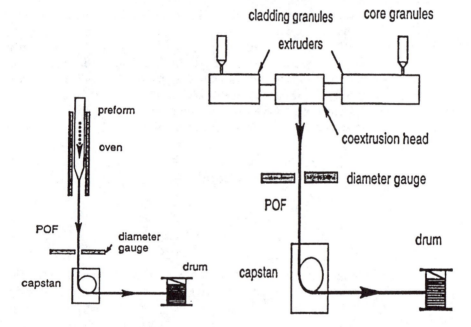

Figure 3. Preform and extrusion apparatus for producing POFs, ref. 6. Reprinted with permission from John Wiley & Sons Limited.

polymerization. Following preform preparation, the bottom of the preform is heated and the fiber is drawn around a capstan. This technique has advantages in its simplicity and in the fact that the polymer is not subjected to prolonged high temperatures or to mechanical degradation processes. However, it is a batch processing technique.

Both batch and continuous coextrusion fiber processing schemes have been developed (13). Figure 3 three illustrates one type of continuous processing scheme. Polymer is fed into separate cladding and core extruders. First, the core material passes through a die; molten cladding material passes through a second die that surrounds the core material. A capstan draws the fiber and transports it to a winding roll. The extrusion process is problematic in that prolonged exposure to high temperatures, high shear stresses and prolonged metal contact renders the polymer labile to degradation. There is also a disadvantage in using polymer that is synthesized external to the extrusion system. Contaminants can enter the polymer during storage and transport. A solution to this dilemma is to connect a batch or continuous polymerization reactor to the extrusion units. In one system, monomers are purified by distillation, fed into the reactor with small amounts of necessary additives, polymerized and coextruded.

Graded-Index Fibers:

The need for high bandwidth, low loss optical fibers has spurred much research on the processing of gradient index plastic optical fibers, GI POFs. Both preform and extrusion techniques have been developed along with novel polymerization schemes employing diffusion mechanisms to produce gradient in refractive indexes. Koike et al extensively studied interfacial polymerization techniques used to produce preforms (16-21). The interfacial polymerization technique involves polymerizing two or more monomers of different refractive indexes inside a tube made of polymer, such as poly (methyl methacrylate) PMMA. When methyl methacrylate (MMA) and vinyl benzoate (18) are polymerized inside the polymer tube, the monomer mixture swells the tube and polymerization proceeds from the wall of the tube inward. The key to producing a gradient in refractive index, which increases from the edge to the center of the rod, lies in designing a system with selected monomer reactivity. If M_1 has a lower refractive index than M_2, and M_1 is more reactive than M_2, the polymer formed at the tube interface has a higher content of M_1 and the M_1 content slowly diminishes as the center of the system is approached. UV and thermal initiation systems have been used. Another system involves a similar approach using a polymer tube, M_1 of low refractive index (MMA) and a non-reactive component of higher refractive index (bromobenzene) (18). Again, the polymerization proceeds from the edge to the center of the tube, leaving a gradient in bromobenzene, which increases, from the edge to the center. The resulting preforms are drawn in the conventional way.

Chen has developed techniques for coextruding GO POFs (22-25). Two supply tanks feed polymer solutions to the extrusion zone. For example, tank 1 contains

PMMA and MMA monomer and benzyl methacrylate monomer (BzMA), while tank 2 contains PMMA and MMA. (The refractive indexes of PMMA and BzMA are 1.490 and 1.568, respectively.) The solutions are heated to 60 °C while gear pumps feed the solutions (tank 1 inner layer and tank 2 outer layer) to a concentric die. A bi-layer composite is extruded and fed to a closed diffusion zone. BzMA diffuses to the outer layer, while MMA diffuses to the inner layer. The material is then fed to an UV curing zone and polymer fiber is collected on a take-up roll.

In all fiber-processing procedures it is of utmost importance to avoid contamination and degradation, since these contribute to optical loss.

Optical Properties and Loss

Light power is absorbed or scattered in the core material or at the core-cladding interface. POFs have historically suffered from high attenuation losses and much research is aimed at altering polymer structure, processing techniques and fiber design to achieve lower loss limits. Current silica based fibers have attenuation limits of 0.3 dB/km (26). However, the purest glass obtainable during the 19th century and up until 1969 had losses of up to 1000 dB/km (27). Just as much research has been devoted to the purification and processing of silica, POF producers are overcoming barriers that have resulted in significant improvements in attenuation. Today, commercially available, PMMA, step index optical fibers exhibit losses of 100 dB/km in the visible region and perfluorinated GI POFs have attenuation of 40 dB/km in the near infra-red region (26). Specific applications require transparency windows in specific regions of the spectrum from the UV/visible to the near infrared region. Factors effecting and methods of reducing these losses are discussed.

Factors Effecting Loss:

Intrinsic Losses: Absorption and Scattering:

Intrinsic losses result from the physical and chemical structure of the polymer. They are due to absorption and scattering. Absorption of light in the ultraviolet, visible and near infrared regions of the electromagnetic spectra is directly related to polymer chemistry. Light induced electronic transitions in polymers are responsible for absorptions in the UV region, which may tail into the visible region depending on the extent to which electrons are delocalized in the structure (28-30). Molecular orbital (MO) theory accounts for electrons taking part in sigma and pi bonds as well as electrons in non bonding energy levels in atoms such as nitrogen, oxygen, sulfur and halogens.

Figure 4 depicts the electronic transitions arising from the placement of these electrons. The transitions increase in energy in the order: $\mathbf{n \Rightarrow \pi^* < \pi \Rightarrow \pi^* < n \Rightarrow \sigma^* <<< \sigma \Rightarrow \sigma^*}$ (26). $\sigma \Rightarrow \sigma^*$ require the most energy and occur at wavelengths as low as 135 nm. Compounds with only sigma bonds are transparent in the UV region of the electromagnetic spectrum. Polymers such as poly (4-methylpentene-1) (PMP) (figure 5) exhibit minimum UV absorption and are attractive candidates for use in fibers, which transmit in this region. PMMA(figure 5) exhibits $\pi \Rightarrow \pi^*$ transitions of low intensity near 200nm. $n \Rightarrow \pi^*$ transitions from the double bond in the ester group result in weak absorptions near 220-230nm (1, 14, 31). At this point in time PMMA is one of the most widely used optical fiber polymers. The phenyl ring in polystyrene is responsible for $\pi \Rightarrow \pi^*$ transitions which absorb light less than 300 nm, but tail into the visible region (14). While polystyrene (figure 5) has a desirable refractive index for core materials, 1.590, PS fibers must be used at wavelengths sufficiently removed from the near visible region. There has been an increasing interest in optical fiber systems that function in the near infrared region (32-41). Recent breakthroughs in POF technology led to the production of perfluorinated plastic fiber that functions in the telecommunications window from 1300 to 1550 nm (42). The fundamental obstacle to overcome in designing fibers for use in the near to mid infrared region is to eliminate higher harmonics of IR absorptions such as those due to C-H stretching. (32-34). The fundamental frequency of the stretching vibration of the C-H system is calculated from (13):

$$\nu_0 = [K/(4\pi^2\mu)]^{1/2}$$

μ is the reduced mass defined as: $\mu = m_1 m_2 / (m_1 + m_2)$

Additional harmonics occur near multiples of the fundamental frequencies, as shown in figure 6. Other atom pairs have fundamental stretching vibrations with higher wavelengths that shift overtone absorptions beyond that of the critical window needed. Table 1. lists these atom pairs. It is evident that replacing carbon-hydrogen bonds with carbon-deuterium or carbon-halogen bonds will result in structures that exhibit increased transparency in the near and mid infrared region. Significant improvements in attenuation losses have been reported for fluorinated and deuterated fibers and waveguides. These are discussed in refs. 26 and 43. Deuterated polymers are, however, costly and difficult to synthesize. Fluorinated polymer structures have provided the most efficient systems to date. Other benefits from fluorinated are discussed in the following sections.

Raleigh scattering is another intrinsic loss mechanism in pure polymers; it occurs from random density fluctuations and anisotropic structure. The order of size of these irregularities is about 1/10 of a wavelength. Loss due to Raleigh scattering, α_s, is expressed as (14):
$$\alpha_s = [C (n^2 - 1)^2 (n^2 + 2)^2 kT\beta_T]/ \lambda^4$$
where, C is a constant, n is the refractive index, β_T is the isothermal compressibility at the temperature, T, and k is the Boltzmann constant. It is evident that losses due to

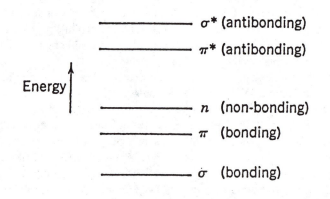

Figure 4. Energy levels for electrons according to MO Theory.

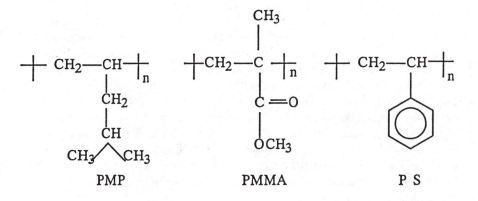

Figure 5. Structures of some common optical polymers.

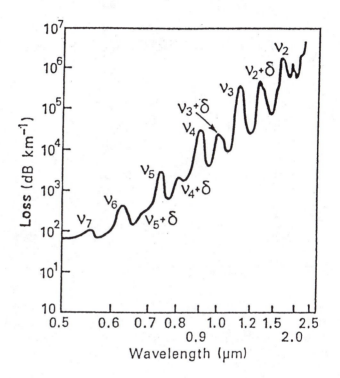

Figure 6. Overtone absorptions in PMMA, ν (stretching) and δ (bending), ref. 13. Reprinted with permission from Kluwer Academic Publishers

Raleigh scattering rapidly diminish as λ increases. Decreasing the refractive index of the polymer also results in diminished scattering. To date, fluorinated core materials with low refractive indexes have produced fibers with the lowest attenuation losses. α_s decreases linearly with temperature until the glass transition temperature, T_g, is reached. The value of β_T at T_g is used in calculating Raleigh scattering , since at T_g the polymer structure is frozen and density fluctuations are inhibited. Raleigh scattering in pure PMMA samples at 633 nm is 9.7 dB/km^{-1} (44).

TABLE 1. λ OF FUNDAMENTAL STRETCHING VIBRATIONS

ATOMIC BOND	λ_0 (nm)
Si-O	9,000-10,000
C-C	7,600 -10,000
C-H	3,300 –3,500
C-D	4,500
C-F	7,600 -10,000
C-Cl	11,700 –18,200
C-O	7,900 -10,000
C=O	5,300 –6,500
O-H	2,800

SOURCE: Reproduced from ref. 39.

Yamashita and Kamada (45) calculated intrinsic losses in polycarbonate (PC) core (bisphenol A type) optical fiber. Table 2 summarizes losses due to electronic transition absorption, ETA, molecular vibration absorption, MVA and wavelength dependent light scattering, LS.

Extrinsic Losses: Absorption and Scattering

Extrinsic losses in optical fibers and waveguides arise from impurities and additives that absorb light as well as from inclusions and core-cladding interface imperfections that scatter light. The earliest POFs had losses of up to 1000 dB/km and this was due mainly to absorption in poorly purified polymer (26). The primary sources of extrinsic absorption in POFs and waveguides are transmission metal contaminants and water. For example, cobalt ions absorb at 530, 590 and 650 nm. It is estimated that 2 ppb concentrations increase attenuation losses up to 10 dB/km (1). Organic contaminants such as initiators and chain transfer agents also contribute to absorption losses. Absorbed water presents a great problem; O-H stretching and bending overtones in the near infrared region are significant enough to effect losses in POFs and in conventional silica fibers. Hida and Imamura (43) tested the influence of humidity on optical waveguide circuits made of deuterated and fluorinated methacrylate polymers. A lost peak at 1410 nm exhibited a linear relationship with humidity due to the second overtone of the O-H stretching vibration.

TABLE 2. INTRINSIC LOSSES IN PC-POF

WAVELENGTH nm	450	500	650	764
ETA (dB/km)	2430	750	64	19
WAVELENGTH nm	650	764	946	
MVA (dB/km)	26	165	1250	
WAVELENGTH nm	488	650	764	
LS (dB/km)	277	78.8	40.9	

SOURCE: Data summarized from ref. 45.

Wavelength independent scattering occurs when dust and microvoids greater that one micron are present (1). Bubbles and cracks cause such scattering as well. Wavelength independent scattering also accompanies orientation-induced changes in refractive index due to fiber processing (12). Core-cladding boundary defects and core diameter variations contribute to extrinsic scattering losses.

Much research behind the development of both silica and plastic optical fibers has been concerned with eliminating and minimizing extrinsic losses. It is interesting to note that short path length waveguide structures with more complicated geometries and production techniques often have very high losses. Such waveguides are used as optical interconnects and in optoelectronic integrated circuits. For example, polymeric, ridge (channel) optical waveguides 15 microns wide and 10 microns thick were fabricated with heat resistant, fluorinated polyimide (46). Losses of 0.7 dB/cm at 630 nm were reported and were highly acceptable for the application. 57-cm long waveguide circuits were produced using deuterated and fluorinated methacrylate polymers; these structures exhibited losses of 0.1 dB/cm at 1,300 nm (43).

Refractive Index:

The refractive index is an important property in determining the light harvesting efficiency in fibers and other waveguides. However, substituents and backbone atoms that control the refractive index also effect Raleigh scattering losses, and mechanical, thermal and surface properties of the system. In this section the relationship between structure and refractive index will be analyzed. Thermal and mechanical properties will be investigated in the following sections.

Bicerano reviews various techniques for predicting properties of polymers via computational modeling (47). Group contribution methods have been widely used for determining the refractive index of polymers. Van Krevelen expands on this topic (48). The molar refraction, R, and molar volume, V, are additive molar quantities used to determine the refractive index, n. Several definitions of molar refraction have been proposed. Three of these definitions correlate well with the refractive indexes of polymers:

1. Lorentz Lorentz method:
$$n = \{ [1 + 2R_{LL}/V]/ [1 - R_{LL}/V] \}^{\frac{1}{2}}$$
2. Gladstone Dale method:
$$n = 1 + R_{GD}/V$$
3. Vogel method:
$$n = R_V/M$$, where M is the molar mass of the monomer unit

The Molar refraction is determined by the interaction of electromagnetic waves with matter. The refractive index increases with R for samples with constant density and increases as the molar volume decreases for samples with constant R. The molar volume increases with temperature and n decreases 2×10^{-4} / K increase in glassy polymers. The refractive index is also wavelength dependent (20), although not to such an extent that there is an appreciable effect on most waveguide performance.

Table 3 lists the refractive indexes of a variety of optical polymers. Two of the trends that are predictable from group contribution methods are:
1. The refractive index decreases with fluorine content.
2. The refractive index increases with chlorine, bromine and phenyl ring content.

TABLE 3. CORE AND CLADDING POLYMERS

Polymers	Refractive Index*	Reference
Aromatic polyphosphazenes	1.61-1.75	49
Poly(pentabromophenyl methacrylate)	1.71	50
Poly(2,6-dichlorostyrene)	1.62	50
Poly(2-chlorostyrene)	1.61	48
Polystyrene	1.59	48
Poly[1,1-ethane bis(4-phenyl) carbonate]	1.59	48
Poly(methyl methacrylate)	1.49	48
Poly(2,2,2-trifluoroethyl methacrylate)	1.42	51
Poly(2,2,3,3-tetrafluoropropyl methacrylate)	1.42	51
Poly(2,2,3,3,3-pentafluoropropyl methacrylate)	1.30	51
Poly(hexafluoroisopropyl methacrylate)	1.39	51
Poly(1H,1H-heptafluorobutyl methacrylate)	1.38	51

* Measurements made at 20 °C with the D line of sodium, λ = 589.3 nm.

Simple calculations predict, for example, that the numerical aperture for step-index polystyrene fiber clad with PMMA is 0.555 as compared to 0.715 for polystyrene clad with poly (2,2,2--trifluoroethyl methacrylate). Fluorinated polyimides containing phenyl rings have, however, been used as core materials in low loss fibers and other waveguides (46). The combination of fluorine and phenyl moieties allows the refractive index of the core materials to achieve a high enough NA when combined with non-phenyl containing, fluorinated cladding materials. Fluorine substituents diminish Raleigh scattering and the polymers exhibit windows of optical transparency in the near infrared region. The fluorine substituents also impart humidity resistance to the materials. Thus, structure-property relations allow one to optimize refractive indexes along with other important fiber properties.

Thermal and Mechanical Properties of Optical Polymers

Although there are many advantages to using POFs, silica fibers can be used in temperature environments that far exceed those of conventional polymers. It is of great interest to identify existing polymers and to design new polymers for use in high temperature environments. Solar technology, sensor technology, automobile data transmission systems and outer space structures are only a few examples of applications for such ruggedized POFs.

There are two main strategies for designing high temperature waveguides systems:

1. Synthesize new, high T_g polymers
2. Cross-link conventional polymer systems.

Most linear, high molecular weight, amorphous polymers form structurally sound members which do not deform under their own weight when they are below their glass transition temperatures. As the glass transition region is approached, however, segmental motion in the polymer backbone results in flow and irreversible deformation. The glass transition is actually a rate dependent phenomenon and measured glass transition temperatures depend on the conditions under which samples are tested. Most tables of glass transition temperatures, unless otherwise specified, report values obtained from differential scanning calorimetry with heating rates at 3-5 °C/min. Glass transition temperatures for some common optical polymers are listed in table 4. The melting point, T_m, of PMP is listed in this table. PMP is the only known semi-crystalline polymer that is transparent in the UV-visible region of the electromagnetic spectrum. Since the crystalline regions pin the amorphous regions, flow is retarded until the melt temperature is reached. PMP, in addition, is one of the most UV transparent polymers. However, anisotropy induced by fiber processing may increase intrinsic scattering losses.

TABLE 4. GLASS TRANSITION TEMPERATURES OF OPTICAL POLYMERS

Polymers	T_g °C	T_m °C
PMMA	105	
PS	100	
PC	150	
PMP	18	300

The glass transition temperature of a given series of polymers can be increased by incorporating stiff structures such as aromatic rings and by increasing hydrogen bonding and other dipole interactions. For example, halogenation has been shown to increase the glass transition temperature in a series of methacrylate and acrylate polymers (52-54). Waveguides produced with fluorinated, aromatic polyimide structures exhibit high thermal stability and optical transparency after being cured above 350 °C (46). The groups that are chosen to enhance thermal properties may also enhance optical properties. Halogen substituents and aromatic groups alter the refractive index and increase near infrared transparency. Fluorination diminishes intrinsic scattering. Other novel, high T_g polymers are reviewed in a later chapter in this symposium series.

Linear (thermoplastic) polymers flow as the glass transition temperature is approached. Cross-linking produces thermosetting polymers that soften, but do not undergo irreversible flow at T_g. Abe et al produced highly heat-resistant acrylic and siloxane polymers by cross-linking (55). Figure 7 compares cross-linked acrylic core material to thermoplastic acrylic and polycarbonate core material. Samples were heated a 1 °C/min up to 200 °C. The cross-linked sample retained over 90% of light transmission efficiency, whereas transmission in linear samples diminished drastically as the T_g was approached. In addition, the cross-linked sample showed 80%/m retention of light after 1,000 hours at 150 °C. Similarly, Takezawa et al (34) produced POFs with cross-linked fluoroalkyl methacrylate cores. These fibers showed similar optical losses to PMMA POFs prior to heat treatment. After heating in air to 160 °C, the fluorinated fibers exhibited losses that were ten times smaller than that of PMMA fibers. Furthermore, the fluorinated fibers had an optical window at 780 nm where there was hardly any change in loss; these fibers are optimum candidates for near infrared optical communications. The mechanism of thermal degradation is of interest. The author's propose that fluorination decreases the formation of conjugated carbonyl bonds formed by thermal oxidation. This

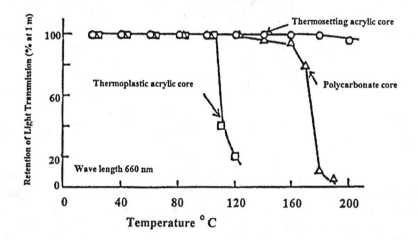

Figure7. Heat-resistance characteristics (short time heating at 1 °/min increase) taken from ref.55 . Reprinted with permission from SAE paper number 910875 © 1991 Society of Automotive Engineers.

illustrates the fact that while cross-linking inhibits attenuation losses due to flow induced imperfections in the fiber geometry, the polymer chemistry controls oxidative and thermal degradation reactions.

Current research devoted to the design of heat resistant linear and cross-linked polymers will undoubtedly lead to many new optical polymers for use in systems that are presently limited to the use of silica fibers.

Optical Fiber Testing

The stringent requirements for plastic optical fibers and optical fibers in general are many, depending upon the application. It is not the intent of this overview to review all of techniques for measuring optical fiber performance. Rather, readers are referred to references 2, 4-6, 15, 27, and 56-60 for information on specific test procedures.

Applications

Telecommunications:

Both step and graded-index POFs are used in telecommunications applications. Conventional, PMMA SI POF is used in short distance systems that require band widths of up to 12 MHz-km. Losses of 0.2 dB/m at 660 nm have been reported for 0.3 numerical aperture fibers (61). PMMA fibers have minimums in attenuation around 520 nm, 570 nm and 650 nm and can be used with a variety of LEDs (62). GI PMMA fibers have bandwidths of 1.25 GHz-km and operate on the Giga bit level with reported losses of 0.15 to 0.8 dB/m at 660 nm (61). 0.5 mm Polycarbonate step-index fibers have transparency windows in the 780 nm region and losses of 0.3 dB/m. These fibers are clad with silicone resins and exhibit thermal resistance up to 125 °C (60). They are designed for optical communication media used in homes and offices. As mentioned earlier, the most promising bandwidth and transmission distances have been achieved by perfluorinated GI POF (26).

Imaging and Illumination:

Light piping is the most basic use for optical fibers; light is carried from one place to another. Light is transported by a single fiber or by bundles of fibers. Step index-fibers or bare fibers are used for these applications. Special POFs have been designed for use in polymer light conduits (PLCs). These devices are generally larger that 3 mm in diameter and are generally made by filling fluoropolymer tubes with liquid monomers which are subsequently polymerized (63). Imaging is a special application of light piping. An image is carried from one end of a bundle to the other; each fiber core captures a section of the image and transmits it to the other end of the fiber (2). Imaging bundles can be placed in areas inaccessible to lenses. The light is then directed to a detector. Endoscopes are examples of this type of imaging. POFs are especially adapted to this application, since they are flexible and only relatively short attenuation lengths are required.

Dye-Doped Optical Fibers:

Solid state, dye-doped materials have many applications in fiber and waveguide optics. Organic polymers are superior to silica in that they are easily doped with organic dyes. Dyes can be incorporated into the polymer matrix during polymerization and after polymerization by solution or compounding processing. Not all dyes, however, survive the polymerization process, so each specific system needs to be evaluated. The temperatures necessary for doping and fabrication of silica systems preclude the use of organic dyes. In addition, organic dyes have very limited solubility in silica.

Some important uses for dye-doped systems are:

1. Luminescent Solar Concentrators (64-69): Polymer fibers, plates or films are doped with fluorescent dyes that absorb sunlight. A series of fluorescent dyes is used in a cascading system. The first dye absorbs the highest energy sunlight and re-emits it at wavelengths where the second dye absorbs. The process is repeated with one or more dyes and the longest wavelength light is absorbed by a photovoltaic device and converted to electricity. The light stability of both the polymers and dyes is critical.

2. Scintillators (70-73): Plastic scintillators are used to detect ionizing radiation. When polymer containing phenyl rings (for example, polystyrene or poly (4-methyl styrene)) comes is contact with high energy radiation, electrons in the phenyl rings are excited. Relaxation to the ground state is accompanied by the emission of photons in the ultraviolet region of the electromagnetic spectrum. Fluorescent dyes incorporated in the polymer matrix absorb the photons and reemit them at longer wavelengths. A detection device collects the emitted photons. In some scintillator a secondary dye is used to shift the wavelength of the photons to lower energies where detection is more efficient. Much research has been devoted to the development of radiation hard dyes and polymers.

3. Sensors: Fluorescent plastic fiber sensors are being developed for use in a variety of sensor devices. For example, a novel system was developed to detect corona discharge in high voltage systems (74). Part of the light emitted by corona discharge is absorbed and re-emitted by the doped fiber core and transmitted to a detector via a transparent optical fiber. Another interesting design comprises a clear polymeric core clad with fluorescent dye-doped polymer used to detect humidity. The emission spectra of the dye changes linearly with the percent humidity (75). Similarly,

fluorescent dye-doped cladding materials are used in POF systems to detect fluorescence quenching in fiber optic oxygen sensors (76).

Conclusions

This overview chapter is designed to introduce polymer chemists to the design and applications of polymeric waveguides and fibers. Many new polymer systems can be designed to expand the efficiency of and the number of applications for polymer waveguides. Graded-index materials, ruggedized systems and dye-doped systems are topics for continued research. Structure-property relations allow chemists to predict and design polymers with minimum intrinsic losses for these uses. Always of interest are new synthetic and processing techniques designed to minimize extrinsic losses.

Acknowledgments

This work was supported by the University of South Florida, the State of Florida and Honeywell, Inc. together via the I-4 Corridor Initiative.

References

1. Kaino, T.; "Polymer Optical Fibers", in *Polymers for Lightwave and Integrated Optics Technology and Applications*; L. A. Hornal, Ed; Marcel Dekker: New York; 1 (1992).
2. Hecht, J.; *Understanding Fiber Optics*; Prentice Hall: Columbus, OH, 1999.
3. Stegeman, G. I., "Waveguiding and Waveguide Applications of Nonlinear Organic Materials", in *Materials for Nonlinear Optics*; S. Mardner, J. Sohn and G. Stucky, Eds; ACS Symposium Series 455, W: Washington, DC, 113 (1991).
4. Shotwell, R. Allen; *An Introduction to Fiber Optics*; Prentice Hall: Columbus, OH, 1997.
5. Sterling, D.; *Designers Guide to Fiber Optics*; AMP Inc.: Harrisburg, PA, 1982.
6. *Plastic Optical Fibres Practical Applications*; Marceau, J., Ed.: John Wiley & Sons, NY, 1997.
7. Shiue, S., *Polym. Eng. And Sci.,* **1998**, 38, 1023.
8. Hayes, J., *Fiber Optics Technician's Manual,"* Delmar Publishers: New York,20(1996).
9. Shitov, V., Chuprakov, V., Nekhorosheva, R., and Kononova, N., *Elektrosvyaz,* **1989**, 8, 37.
10. Shiue, S. and Lee, S., *J. Appl. Phys,* **1992**, 1, 18.
11. Theis, T., Brockmeyer, A., Groh, W., and Stehlinn T. F..; "Polymer Optical Fibers in Data Communications and Sensor Applications", in *Polymers for*

Lightwave and Integrated Optics Technology and Applications; L. A. Hornal, Ed; Marcel Dekker: New York; 39 (1992).

12. Kaino, T., Fujiki, M., and Nara, S., *J. Apply. Phys.*, **1981**, 52, 7061.

13. Emslie, C., *J. of Materials Sci.*, **1988**, 23, 2281.

14. Glen, R, M., *Chemtronics*, **1986**, 1, 98.

15. Yeh, C.; *Handbook of Fiber Optics Theory and Applications;* Academic Press, NY., 25, 1990.

16. Ishigure, T., Nihei, E., Yamazaki, S., Kobayashi, K. and Koike, Y., *Electronics Letters*, **1995**, 31, 467.

17. Koike, Y., *Polymer*, **1991**, 32, 1737.

18. Ishigure, T., Nihei, E., and Koike, Y., *Applied Optics*, **1994**, 33, 4261.

19 .Kobayashi, T., Tagaya, A., Nagatsuka, S., Teramoto, S., Nihei, E., Sasaki, K., and Koike, Y., "High-Power Optical Fiber Amplifiers in the Visible Region," in *Photonic and Photoelectric Polymers*, S. A. Jenekhe and K. J. Wynne, Eds. ACS Symposium Series 672: Washington, DC. 47 (1995).

20. Nihei, E., Ishigure, T., and Koike, Y., " Optimization and Material Dispersion in High-Bandwidth Graded-Index Polymer Optical Fibers," in *Photonic and Photoelectric Polymers,* S. A. Jenekhe and K. J. Wynne, Eds; ACS Symposium Series 672: Washington, DC.; 58 (1995).

21. Koike, Y., "Graded Index Materials and Components," in *Polymers for Lightwave and Integrated Optics Technology and Applications*; L. A. Hornal, Ed; Marcel Dekker: New York; 71 (1992).

22. Chen, W. C., Chen, J. H., Yang, S. Y., Chen, J. J., Chang, Y.H., Ho, B. C., and Tseng, T. W., " Preparation and Characterization of Gradient-Index Polymer Fibers," in *Photonic and Photoelectric Polymers,* S. A. Jenekhe and K. J. Wynne, Eds. ACS Symposium Series 672: Washington, DC.; 71 (1995).

23. Yang, S. Y., Chang, Y. H., Ho, W. C., Chen, W. C., and Tseng, T. W., *Polymer Bulletin*, **1995**, 34, 87.

24. Ho, B. C., Chen, J. H., Chen, W. C., Chang, Y. H., Yang, S. Y., Chen J. J., and Tseng, T. W., *Polymer Journal, The Soc. Polym. Sci., Japan*, **1995**, 27, 310.

25. Chen, W. C., Chen, J. J., Yang, S. Y., Cheng, J. Y. Chang, Y. H., and Ho, B. C., *J. Apply. Poly. Sci.*, **1996**, 60, 1379.

26. Koike, Y. and Ishigure, T., *IEICE Trans.Commun.*, **1999**, 82-B, 1287.

27. Einarsson, G.: *Principles of Lightwave Communications:* John Wiley and Sons: NY, 5 (1996).

28. Braum, A. M., Maurette, M. T. and Oliveros, E., *Photochemical Technology*, New York: John Wiley and Sons, pp. 1-50 (1991).

29. Williams, D. H., and Fleming, I., *Spectroscopic Methods in Organic Chemistry,* 5th Ed., New York: McGraw-Hill, pp. 1-27 (1995).

30. Pecsok, R. L., Shields, D., Cairns, T., and McWilliam, I. G., *Modern Methods of Chemical Analysis,* New York: John Wiley and Sons, pp. 226-242 (1976).

31 .Lippet, T., Webb, R. l., Langford, S. C., and Dickinson, J. T., *J. Apply. Phys.* **1999**, 85, 1838.

32. Takezawa, Y., Taketani, N., Tanno, S., and Ohara, S., *J. Polym. Sci.: Part B: Polym. Phys,* **1992**, 30, 879.
33. Drexhage, M. G., and Moynihan, C. T., *Scientific American,* **1988**, November, 110.
34. Takezawa, Y., Tanno, S.,Taketani, N., Ohara, S., and Asano, H., *J. Appl. Poly. Sci.,* **1991**, 41, 3195.
35. Groh, W., *Macromol. Chem.,* **1988**, 189, 286.
36. Ishigure, T., Nihei, E., Koike, Y., Forbes, C., LaNieve, L., Straff, R., Deckers, H. A., *IEEE Photon. Tech. Letts,* **1995,** 7, No. 4, 403.
37. Takezawa, Y. and Ohara, S., *J. Appl. Poly. Sci.,* **1993**, 49, 169.
38. Kalish, D., and Clayton, J., " Challenges for POF in Premises Network", in *Proceedings POF Conference '97,* Hyatt Regency Kauai Resort & Spa, Kauai, Hawaii, September 22-25, p. 1 (1997).
39. Yoshihara, N., "Performance of Pefluorinated POF", in *Proceedings POF Conference '97,* Hyatt Regency Kauai Resort & Spa, Kauai, Hawaii, September 22-25, p. 27 (1997).
40. Tanio, N. and Koike, Y., " What is the Most Transparent Polymer?", in *Proceedings POF Conference '97,* Hyatt Regency Kauai Resort & Spa, Kauai, Hawaii, September 22-25, p. 33 (1997).
41. Imai, H.,"Applications of Perfluorinated Polymer Fibers to Optical Transmission", in *Proceedings POF Conference '97,* Hyatt Regency Kauai Resort & Spa, Kauai, Hawaii, September 22-25, p. 29 (1997).
42. *Harnessing Light Optical Science and Engineering for the 21st Century,* National Research Council, Washington, DC: National Academy Press, 255 (1998).
43. Hida, Y., and Imamura, S., *Jpn. J. Appl. Phys.,* **1995**, 34, 6416.
44. Tanio, N., Kato, H., Koike, Y., Bair, H. E., Matsuoka, S., and Blyler, Jr., L. L., *Polymer J. 1998,* 30, 56.
45. Yamashita, T., and Kamada, K., *Jpn. J. Appl. Phys.* **1993**, 32, 2681.
46. Matsuura, T., Ando, S., Sasaki, S., and Yamamoto, F., *Electronice Letters,* **1993**, 29, 269.
47. *Computational Modelling of Polymers;* Bicerano, J., Ed.; Marcel Dekker; New York: 1992, 115.
48. Van Krevelen, D.W., *Properties of Polymers*; Elsevier: New York, 1990, 292.
49. Olshavsky, M. A. and Allcock, H. R., *Macromolecules,* **1995**, 28, 6188.
50. Brandrup, J. and Immergut, E. H., *Polymer Handbook,*Wiley Interscience: New York, 1989, 457.
51. Gaynor, J., Schueneman, G., Schuman, P. and Harmon, J. P. *J. Apply. Polym. Sci.,* **1993**, 50, 1645.
52. Gaynor, J., Schueneman, G., Schuman, P., and Harmon, J., *J. Apply. Polym. Sci.,* **1993**, 50, 1645.
53. Bertolucci, P., and Harmon, J.,"Dipole-dipole Interactions in Controlled Refractive Index Polymers," in *Photonic and Photoelectric Polymers,* S. A. Jenekhe and K. J. Wynne, Eds; ACS Symposium Series 672: Washington, DC.;791 (1995).

54. Bertolucci, R., Harmon, J., Biagtan, E., Schueneman, G., and Goldberg, E., *Poly. Eng. And Sci.,* **1998**, 38, 699.

55. Abe, T., Asano, H., Okino, K., Taketani, N., and Sasayama, T., *S. A. E. Transactions,* Society of Automotive Engineers: New York City, 1991, 1298.

56. Cohen, L., Kaiser, P., Lazay, P., and Presby, H., "Fiber Characterization" in *Optical Fiber Telecommunications,* S. E. Miller and A. Chynoweth, Eds., Academic Press : New York; 343 (1979).

57. Marcuse, D., *Principles of Optical Fiber Measurement,* Academic Press: New York (1981).

58. Stewart, W. J., *J. of Quantum Electronics,* **1982**, QE-18, 10, 1451.

59. Blyler, L. L., Salamon, T., White, W. R., Dueser, M., Reed, W. A., Koeppen, C., Wiltzius, P., and Quan, X., *Int. Cable and Wire Proceed.,* **1998**, 241.

60. Hattori, M., Nishiguchi, M., and Takagi, S. ., *Int. Cable and Wire Proceed.,* **1998**, 257.

61. Nishiguchi, M., Hattori, M., and Takagi, S., *Int. Cable and Wire Proceed.,* **1998**, 248.

62. Ziemann, O., Ritter, T., and Gorzitza, B., *Int. Cable and Wire Proceed.,* **1998**, 264.

63. Che., W., and Chang, C., *J. Mater. Chem.,* **1999**, 9, 2307.

64. Salem, A. i., Mansour, A. F., El-Sayed, N. M., and Bassyouni, A. H., *Renewable Energy,* **2000**, 20, 95.

65. Bakr, N., Mansour, A., and Hammam, M., *J. Apply. Polym. Sci.,* **1999**, 74, 3316.

66. Mansour, A., Salem, A., El-Sayed, N., and Bassyouni, A., *Proc. Of Indian Acad. Of Sci: Chem. Sci.,* **1998**, 110, 351.

67. El-Shahawy, M., and Mansour, A., *J. Mater. Sci.,* **1996**, 7, 171.

68. Sakuta, K., Sawata, S., and Tanimoto, M., *IEEE Proc. Of Photovoltaic Specialists Conference, Part 1, Waikoloa, HI,* **1994**, 1, 1115.

69. Reisfeld, R., Shamrakov, D., and Jorgensen, C., *Solar Energy Materials and Solar Cells,* **1994**, 33, 417.

70. Bertolucci, P., Harmon, J., Biagtan, E., Schueneman, G., Goldberg, E., Schuman, P. and Schuman, *Polym. Eng. Sci.,* **1998**, 38, 699.

71. Biagtan, E., Goldberg, E., Stephens, R., Valeroso, E., and Harmon, J., *Nucl. Instr. Meth. In Phys. Res.* **1996**, B114, 88.

72. Biagtan, E., Goldberg, E., Stephens, R., and Harmon J., *Nucl. Instr. Meth. In Phys. Res.* **1996**, B114, 302.

73. Biagtan, E., Goldberg, E., Stephens, R., and Harmon, J., *Nucl. Instr. Meth. In Phys. Res.* **1994**, B93, 296.

74. Farenc, J., Mangeret, R., Boulanger, A., and Destruel, P., *Rev. Sci, Instrum,* **1994**, 65, 155.

75. Muto, S., Sato, H., and Hosaka, T., *Jpn. J. Appl. Phys.,* **1994**, 33, 6060.

76. Morisawa, M., Vishnoi, G., Hosaka, T., and Muto, S., *Jpn. J. Apply. Phys.* **1998**, 37, 4620.

Chapter 2

Plastic Optical Fibers: Pipe-Dream or Reality?

Xina Quan, Lee Blyler, and Whitney White

Bell Laboratories, Lucent Technologies, 700 Mountain Avenue, Room 7E–217, Murray Hill, NJ 07974

There has been interest in plastic optical fibers (POF) since the advent of optical fiber communications. The dream has been to build extremely low-cost optical networks from POF manufactured with inexpensive, continuous extrusion processes. Large-diameter POF can be used due to the low elastic moduli of polymers. This relaxes the alignment tolerances needed in these networks which can dramatically decrease overall system costs since simple injection-molded connectors can be used. Installation costs are also expected to be substantially lower, due to simpler end-face preparation techniques. To balance all these potential advantages, there have been a number of issues preventing the widespread adoption of POF. However, recent research results offer the promise of finally overcoming some of these obstacles.

Why are Plastic Optical Fibers so exciting?

The simple answer is "Cost, Cost, Cost". Unlike glass fiber, which is drawn in a batch process, plastic fibers can be extruded in a continuous process at high rates. For simple fiber designs, this is intrinsically a less expensive production technique with relatively low capital costs. A typical layout is shown conceptually in Figure 1. Two different types of polymers or polymer blends are melted and pumped through a co-extrusion die crosshead to make concentric layers for the core and cladding of the fiber. The output of the die is drawn down to adjust the outer diameter of the fiber. After cooling, the fiber is collected continuously on a take-up reel. In theory, plastic optical fiber can be produced as long as plastic pellets are fed into the hoppers of the extruders above.

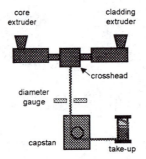

Figure 1. Schematic of a coextrusion system used to produce plastic optical fiber.

The costs of a POF system can also be lower than its glass fiber analog due to connectorization and installation advantages. Most POF designs incorporate extremely large fiber sizes, with optical cores on the order of 500 to 1000 μ diameter. The high modulus of silica limits the diameter of most glass fiber designs to an outer diameter of 125 μ for flexibility. The relative core sizes of glass and plastic multimode fiber are compared graphically in Figure 2.

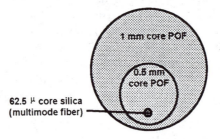

Figure 2. Relative core sizes of glass and plastic multimode fibers.

The major advantage of larger diameter fibers comes from a potential reduction in the coupling loss that is added to an optical network whenever two fibers are connected together. This loss scales with the ratio of the lateral displacement between fibers d to the radius of the fiber core a. Models indicate that to achieve a coupling loss less than 0.5 dB, this ratio d/a should be less than 0.25. For multimode fibers with a core diameter of 62.5 μ, this means the absolute displacement must be less than 8 μ. By contrast, the misalignment in fibers with a 300 μ core can be as large as 38 μ. As can be seen in Figure 3, the much greater displacement allowed for larger fibers significantly relaxes the tolerances required in the alignment process. This, in turn, reduces the costs of attaching fibers to devices

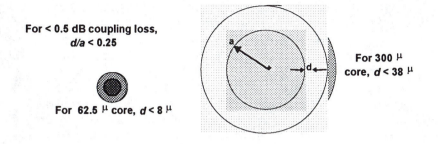

Figure 3. Relative displacements allowed for 0.5 dB connection of glass and plastic multimode fibers with 62.5 μ and 300 μ cores, respectively.

in network endpoints and allows the use of inexpensive, molded connectors that do not require the expensive precision ferrules needed in most connectors used for glass optical fibers.

Another potential advantage that arises from the larger diameter of POF, as well as from the thermoplastic nature of the polymers used, is ease of handling and end-face preparation. A common demonstration with POF is to cut the fiber with a pair of cutters and flatten the ends by touching them to a hot plate. This is clearly much less labor intensive than the strip-cleave-and-polish process needed to prepare glass fiber ends. Melt splicing of POF is also much easier and requires less specialized equipment than high temperature fusion splicing of glass fibers. Since much of the cost of an optical network lies in the installation and maintenance of the network, these attributes of POF are very appealing, particularly for architectures that have many interconnections such as local area networks.

So why hasn't there been widespread use of POF?

Despite intense interest in the technology, the performance of POF has not been adequate for many of the proposed applications such as data networking and equipment interconnection. Until recently, poly(methylmethacrylate) or PMMA has been the material of choice for the optical cores of a plastic fiber. Thus, optical attenuation, or loss, has been high, on the order of 150 dB/km, which greatly restricted the length over which signals could be carried. The transmission capacity, or bandwidth, of the fibers has been limited due to the step index fiber designs available in commercial fibers. In addition, PMMA transmits best at 650 nm or shorter wavelengths for which there are very few practical high-speed sources. This wavelength restriction creates difficulties in using these fibers with components available for communications systems designed for conventional glass fibers that use 850, 1300, and 1500 nm sources.

There have also been difficulties in achieving the expected cost advantages of POF systems. Since they cannot build on the existing infrastructure of glass optical fiber networks, the technology has remained near the beginning of the learning curve and has not benefited from the cost reductions normally experienced with increased production volumes. Installation costs have also been higher than expected since coupling losses generally are high – as high as 2 dB or more – unless the fiber ends are well polished, obviating the labor cost differential of connectorizing plastic and glass fibers. These high coupling losses were a major surprise since the losses of POF are expected to be less than 0.5 dB based on geometrical tolerances. With these considerations, copper cable has been a lower cost alternative for most applications that fall within the bandwidth and distance limitations of commercial POF.

Finally, the long-term and thermal stability of POF has been a major concern. The performance of most polymers at elevated temperatures is clearly less than that of glass or copper, influencing many potential users to choose the more established technologies.

But the picture is changing….

Over the past several years, there have been a number of major advances in the development of POF. New materials and fiber designs have greatly improved both the bandwidth and loss of plastic optical fibers.

Processes have been developed to produce graded-index fibers that have greatly enhanced transmission capacities.[1,2] Traditionally, step-index POF has been produced with one material (generally PMMA) in the light-guiding portion or core of the fiber and an outer coating or cladding of a second material with a lower refractive index (often a fluorinated polymer). This structure is shown schematically in Figure 4. The low bandwidth of step-index fibers can be understood conceptually

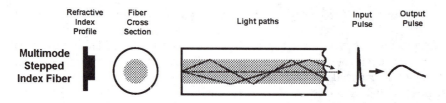

Figure 4. Schematic of light rays traveling through a step-index fiber.

by tracing the rays of light going through the fiber as shown in the figure. If light is launched from a divergent source into the core, a small fraction of the light will go straight down the middle. The rest of the light enters the core at a wide range of angles. It will be reflected at the core-cladding interface if the angle of incidence is smaller than the critical angle of total internal reflection, which is determined by the difference in refractive indices of the core and cladding. Light that comes in at relatively steep angles will undergo a larger number of reflections and travel a greater distance as it moves from one end of the fiber to another. Since the speed of light is constant through the core region, this light will arrive at the receiving end of the fiber after the light that travels down the middle of the fiber. Thus, a detector at the receiving end would record a distribution of arrival times for a single pulsed signal. The broader this distribution, the lower the transmission capacity of the fiber since signal pulses have to be spaced further apart in time to avoid overlap between them.

In a graded-index fiber, the refractive index is modulated throughout the core as shown in Figure 5. In this case, the speed of light is not constant throughout the

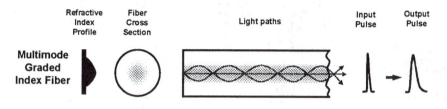

Figure 5. Schematic of light rays passing through a graded-index fiber.

core region. Light travels faster near the core-cladding interface than at the core. With appropriate index profiles, the extra distance traveled by rays launched at steep angles is compensated by their higher velocities and all of the light arrives at the receiving end at about the same time. This narrower distribution of arrival times corresponds to higher bandwidths for the graded-index fiber.

Y. Koike (Keio University) advanced POF technology substantially when he developed methods for fabricating graded-index fiber. One method, interfacial gel polymerization,[2] creates preforms by polymerizing a monomer/dopant blend while it is confined in a glassy polymer tube. As illustrated in Figure 6, some of the dopant molecules get excluded from the polymer gel phase that forms by swelling the polymer tube with monomer. This creates a concentration gradient of dopant molecules, which translates into a refractive index gradient.

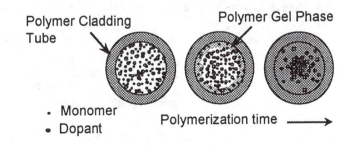

Figure 6. Schematic of Koike's interfacial gel polymerization process to fabricate graded-index POF.

Another approach pioneered by Koike and developed further by the Asahi Glass Company[3] fabricates preforms by simple diffusion of a low molecular weight dopant confined in a glassy polymer tube. Both processes produce fiber that has a graded-index profile and a higher bandwidth than the commercial step-index plastic optical fibers.

Koike has also shown that the loss of the fiber can be greatly improved when it is made from glassy, perfluorinated polymer matrices.[4] These materials do not have the C-H overtones that have caused the high optical attenuation of previous POF as shown in Figure 7.

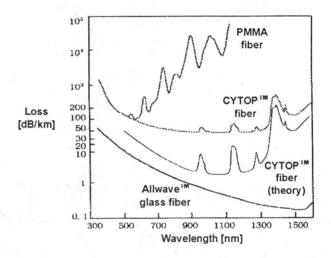

Figure 7. Comparison of optical attenuation of glass, PMMA, and perfluorinated fibers. [Adapted from Reference 7]

With losses now lower than 50 dB/km,[5] signals can be transmitted at higher speeds and over longer lengths of perfluorinated fiber than with POF made from conventional polymer materials. The transmission window of perfluorinated POF is also much broader, and allows the use of wavelengths from 850 to 1300 nm,[7] the standards wavelengths of data and telecommunications networks. Recent results show multiple wavelength windows where multi-gigabit transmission is possible for perfluorinated fibers.[6-8] For example, 11 Gb/s transmission at 1300 nm is possible over 100 m of graded-index, perfluorinated POF.[7] Transmission rates of 9 Gb/s have been demonstrated over 100 m at 850 and 980 nm as well.[8]

A feature that may be unique to POF is an unusually high level of energy transfer between transmission mode groups.[9,10] This leads to several advantages for POF systems. The dependence of bandwidth on the details of the refractive index profile of the fiber core is greatly reduced.[11] This is a major difference from glass multimode fiber, whose bandwidth is very sensitive to small changes in the shape of the index profile.[4] In addition, coupling between many modes within the fiber core ensures that power launched over a large fraction of the core cross-section travels at substantially uniform velocity, greatly increasing the bandwidth.[10] This also simplifies the coupling between fibers and between the fiber and a source or detector since details of the signal launch conditions become much less important.

Coupling losses have benefited from other recent advances as well. We recently discovered that the anomalously high coupling losses in POF were due to deep sub-surface cracks that extended tens of microns from the cut surface of a fiber.[12] Figure 8a shows the extent of these cracks 60 microns below the surface. Clearly, coupling

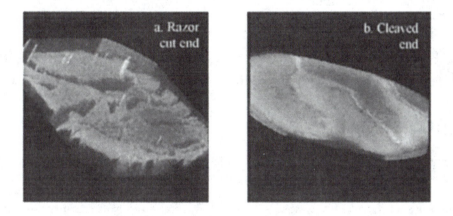

Figure8a. Confocal micrographs showing reconstructed topology of a 1 micron diameter region for a) the first 60 microns below the surface of the fiber end for a fiber cut with a fresh razor blade and b) 30 microns from the end of a fiber cleaved using a new fiber termination process.

losses due to scattering from the cracks will be very high. Polishing the ends improves the situation only after a substantial amount of material is removed. Smoothing the surface by "hot-plate polishing" will not necessarily improve the coupling losses since it simply seals the ends of the cracks, leaving microvoids trapped in the bulk of the fiber.

To address these issues, we have developed fast, inexpensive techniques to prepare the end faces of the fiber.[12] Preliminary results for fiber connections prepared in this way have average coupling losses less than 1 dB. The confocal micrograph in Figure 8b shows the smooth, crack-free surface that can be obtained with one of the new techniques. With these improved connections, the portion of the loss budget of the optical link that is allocated to the fiber can be increased, potentially allowing the use of longer fiber spans or lower cost fiber.

What remains to be done?

Clearly much progress has been made in developing POF fibers that are suitable for communications applications. With graded-index perfluorinated polymer fiber, high bandwidth systems can be designed to run at standard frequencies over span lengths meeting the specifications for local area and home networks. The dream of low-cost, low-loss connectors and end-points appears to be achievable. However, much more work remains to really make POF a success. Three major areas are optical loss, fiber material costs, and long-term reliability.

Optical loss

Although major strides have been made in reducing the optical attenuation of POF, glass fibers are still better by more than an order of magnitude. Bulk scattering studies imply that, although the pure polymer matrix might be expected to have an intrinsic loss of 1 dB or less, the addition of low molecular weight dopants to control the refractive index of the fiber can raise the intrinsic losses to 5-10 dB/km at 850 nm.[13] Extrinsic losses such as processing flaws or compositional variations can add an additional 20 dB/km or more. It is reasonable to expect, however, that these extrinsic losses will decrease as fiber manufacturing processes improve. Glass fiber showed a similar learning curve.

Reducing the intrinsic losses will require additional research on new dopant/matrix systems. It is generally believed that most of these losses are due to scattering from local compositional fluctuations. To reduce them, the compatibility of the dopant molecule in the matrix must be improved. The dilemma is that, in the absence of specific interactions, improved miscibility is best accomplished by increasing the chemical similarity of the two components. This counteracts the desired effect of the dopant molecule to maximize the difference in refractive index between the doped and neat matrix materials. Thus the challenge is to develop systems where the components are highly miscible but have widely different refractive indices. It is particularly difficult to accomplish this in perfluorinated systems. Deuterated and other non-protonated systems have also been studied but have generally been rejected due to cost or long-term reliability considerations.

Fiber material costs

Many of the enhanced properties of the new POF designs are due to the use of amorphous, perfluorinated polymers and dopants. These are substantially more expensive than the materials used in commercial POF, with bulk costs on the order of dollars per *gram*, rather than dollars per *pound*. The challenge of using these materials is to adjust the fiber diameter to balance a reduction in material costs with the added costs of tightening the tolerances of connections in the network. Discovery of alternative low-cost material systems that have the performance of the perfluorinated systems but costs similar to commercial glass or PMMA fiber would greatly improve the probability of widespread adoption of the POF technology.

Reliability

Perhaps the major barrier to success for POF is the perception that plastic products do not have the long-term stability required for communications systems. Adoption of commercial step-index fibers has long been hindered by this perception. The situation has been exacerbated by the development of graded-index fibers where the refractive index profile is modulated by the addition of low molecular weight molecules. The properties of these dopant molecules are chosen to maximize both the change in refractive index and their solubility in the polymer matrix. As discussed above, these properties are often mutually exclusive. Since most fabrication processes call for the diffusion of the dopant molecules into the matrix to form the index profile, the dopants are also chosen for high diffusivity in the molten polymer. However, this can lead to degradation of the performance of the fiber with time or temperature since long-term diffusion can occur within the glassy fiber. Major changes in the index profile under accelerated aging conditions have been observed and fitted with a model for non-Fickian diffusion, although these changes have not yet been correlated to the optical performance of the fiber.[14] Example results are shown in Figure 9. The effect of thermal history on the fiber may be even more pronounced since the addition of dopants often reduces the glass transition temperature of a polymer matrix.

The challenge for POF researchers then is to develop systems where the diffusion of dopant is very high during processing but extremely low under the conditions where the fiber is used or stored. The most attractive route may be to use a polymer matirx with a high glass transition temperature, but this complicates the fiber manufacturing process. The concentrations of dopant needed must be sufficient to raise the refractive index of the core by several percent but cannot be so high that the glass transition temperature of the core is significantly reduced. Accurate models will then need to be developed which can definitively support the long-term stability of fibers fabricated from these materials.

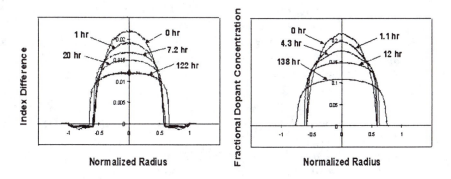

Figure 9. Index profile changes observed (left) and predicted (right) for benzyl benzoate-doped PMMA fibers aged at 109C and 120C, respectively. (adapted from Reference 14)

Conclusions

There have been major advances in the development of POF for communications applications in the last 5 years. It is possible to envision high speed POF networks that are easy enough to connectorize for use in home and office applications. However, further work is still necessary to decrease the optical attenuation, decrease bulk material costs, and increase fiber reliability before POF can make appreciable inroads against the established technologies of copper twisted pair and glass optical fiber.

Acknowledgments

Many of the conclusions in this paper are based on the data generated by other colleagues researching the capabilities of POF at Bell Labs, particularly Art Hart, Todd Salamon, Giorgio Giaretta, George Shevchuk and John Alonzo. Also, Prof. Koike and the members of the NB-1 team at Asahi Glass Company have kindly shared some of their results.

References

1. Ishigure, T.; Nihei, E.; Koike, Y. *Appl. Opt.* **1994**, *33*, 4261.
2. Ishigure, T.; Horibe, A.; Nihei, E.; Koike, Y. *J. Lightwave Technol.* **1995**, *13*, 1475.
3. Yoshihara, N. *Proc. OFC '98, OSA Technical Digest Series Vol. 2*, Optical Society of America, Washington, DC, **1998**, p.308.
4. Nihei, E.; Ishigure, T.; Tanio, N.; Koike, Y. *IEICE Trans. Electron.* **1997**, *E80-C*, 117.
5. Onishi, T.; Murofushi, H.; Watanabe, Y.; Takano, Y.; Yoshida, R.; Naritomi, M. *Proc. POF '98 7th Intl. Plastic Optical Fibres Conference '98* **1998**, p. 39.
6. Imai, H. *Proc. POF Conference '97* **1997**, p. 29. Li, W.; Khoe, G. D.; v.d. Boom, H. P. A.; Yabre, G.; de Waardt, H.; Koike, Y.; Naritomi, M.; Yoshihara, N.; Yamazaki, S. *Proc. POF '98 7th Intl. Plastic Optical Fibres Conference '98* **1998**, *Unpublished and Post-deadline Papers.*
7. Sugiyama, N. *Polymer Preprints, Vol. 39, No. 2*, Division of Polymer Chemistry, American Chemical Society, Washington, DC, **1998**, p.1028.
8. Giaretta, G.; Mederer, F.; Michalzik, R.; et. al. *Proc. ECOC '99*, **1999**.
9. Garito, A. F.; Wang, J.; Gao, R. *Science* **1998**, *281*, 962..
10. Dueser, M.; White, W. R.; Reed, W. A.; Onishi, T. *Technical Digest, Symposium on Optical Fiber Measurements* **1998**, *NIST Special Publication 930*, 135.
11. Yoshihara, N. *Proc. POF Conference '97* **1997**, p. 27.
12. Shevchuk, et al. *Proc. POF Conference '97* Post-deadline paper.
13. White, W.; Wiltzius, P.; Dueser, M. *Proc. 1998 IEEE/LEOS Summer Topical Meeting on Organic Optics and Optoelectronics* **1998**, p.17.
14. L. L. Blyler, Jr., L. L.; Salamon, T.; Ronaghan, C.; Koeppen, C. S. *MRS Symp. Proc. Ser.*, **1998**, *531*, 107.

Chapter 3

Synthesis and Properties of Optically Active Polyurethane Ionomers Containing Erbium

Quan Gu and William M. Risen, Jr.*

Department of Chemistry,Brown University, 324 Brook Street, Providence, RI 02912

Optically active polyurethane compounds in the form of the acid-containing polymers and their ionomers containing erbium and related lanthanide ions have been synthesized. The compounds contain tartaric acid precursors and cyclic trimeric isocyanates as well as relatively long chain diols. The diols employed are based on polybutadiene or carboxylated polyesters. The polyurethanes containing polybutadiene were synthesized using several reaction conditions in order to vary the overall morphology kinetically. The optical activity of solutions of the acid-form polyurethanes and of their erbium ionomers is shown to vary strongly as a function of the time of the first step of the reaction. On a monomer basis, the optical activities are substantially different from that of the tartaric acid precursor. The near infrared and fluorescence spectra of lanthanide-containing polyurethanes are reported.

The goal of this work is to synthesize optically active, acid-form polyurethane polymers that can be converted into the lanthanide ionomer form and used as coatings for optical fibers and related optical structures. Such materials could impart a range of interesting properties, including the ability to rotate the plane of polarization of any photons emitted by the incorporated lanthanides, after their excitation, in order to use the polarization-selective reflection and transmission of light at the polyurethane-substrate interface to advantage. It also would be useful to have an optically active waveguide material that can exhibit useful magnetooptic phenomena such as Faraday rotation. They could have additionally valuable properties if they could be applied as liquids or solutions and photocrosslinked after initial application, so the synthetic goal includes the incorporation of photocrosslinkable units in polyurethane ionomer. Finally, if it had excellent toughness and similar mechanical properties, it could find additional applications.

Synthetic optically active polymers, defined here as those which rotate the plane of polarization of light, have been explored extensively. In some cases, the focus has been on the relation between the helical conformation of the linear polymer and its optical activity (*1 – 4*). In other cases, the optical activity arises from the properties of the groups attached to the polymer chain (*5*). Other types of optically active polymers are ones in which the property of interest is the hyperpolarizability so that nonlinear optical effects can be achieved.

The syntheses of a range of polyurethane ionomers is known and has been reviewed by Xiao and Frisch (*6*). The syntheses of optically active polyurethane acid copolymers which contain photocrosslinkable segments and their erbium ionomers is reported here. These polyurethane acid copolymers were synthesized by combining a cyclic triisocyanate with optically active forms of tartaric acid (a diol) and polybutadiene diol as chain extenders. In order to reach the polyurethane stoichiometry, the total number of isocyanate groups is essentially the same as the total number of alcohol groups on the diols. The reaction conditions were varied systematically in order to explore the strong influence they can exert over the optical rotatory properties of the products. Specifically, the design of the synthesis is to prepare a prepolymer by combining the cyclic trimeric isocyanate with the long chain, non optically active polybutadiene diol for a specific time before adding the tartaric acid. The time of this initial prepolymer synthetic step is then varied to discover its effect on the properties of the polyurethane products.

Experimental Section

Materials.

The materials employed in the syntheses are L-tartaric acid (99.5%), D-tartaric acid (99%), D,L-tartaric acid (99%), ethyl acetate (99.5+%), N,N-dimethylacetamide (99+%, spectro-photometric grade), and dibutyltin dilaurate (95%) from Aldrich Chemical Co.; hydroxyl terminated polybutadiene (hydroxyl value: 1.2 meq/g) from Polyscience Inc.; Desmodur N-3300 or t-HDI (ring form trimer polyisocyanate based on hexamethylene diisocyanate with isocyanate equivalent weight = 195) from Bayer Co.; chloroform from Mallinckrodt Baker Inc.; and erbium acetylacetonate hydrate and terbium acetylacetonate hydrate (99.9%) from Strem Chemicals.

Synthetic procedure.

All of the polymer syntheses were carried out by a two stage process. For those with D- or L-tartaric acid, polybutadiene diol, and N-3300; 0.890 g N-3300,

0.200 g tartaric acid and 1.508 g polybutadiene diol were reacted. This corresponds to about 1.7% overindexing of isocyanate. In the first step the N-3300, polybutadiene and 38.0 ml of ethyl acetate were combined. In the second step, the tartaric acid and 20.0 ml each of ethyl acetate and N,N-dimethylacetamide were added. In each reaction, about 0.1% dibutyl tin dilaurate was used. These materials were reacted for various times in the first step. They are designated by a symbol indicating the type of optically active tartaric acid and the time (t_1) in minutes for the first step. Thus, L90, is the one made with L-tartaric acid and refluxed 90 minutes in Stage 1.

At the beginning of the first stage, all of the long chain diol (polybutadiene diol) and all of the N-3300 required were dissolved in ethyl acetate with a catalytic amount of dibutyltin dilaurate. After stirring for one minute, one drop of the solution was taken for FTIR measurement. The rest of the solution was refluxed for certain amount of time, t_1, at 79°C with stirring under nitrogen protection. The resulting solution was cooled in ice water.

In the second stage of the synthesis, the required amount of short chain diol (tartaric acid or 2,2-bispropionic acid) was dissolved in N,N-dimethylacetamide, and added together with more ethyl acetate to the solution from the first stage. Then the solution was refluxed at 82°C for 43 hours with stirring under nitrogen. The solution was cooled in the air. The ratio of NCO/OH was fixed at 1.02. All the solutions were clear and colorless. One drop of solution was taken for FTIR measurement. These two stages of the process are represented graphically in Figure 1.

Syntheses of comparative compositions were carried out analogously with all possible forms of tartaric acid as short chain diols and with polybutadiene diol as long chain diol.

Preparation of Er^{3+} and related lanthanide ionomer solutions.

The amount of erbium acetylacetonate needed to combine stoichiometrically was calculated by fixing Er^{3+}/COOH to 1:3. This Er(acac)$_3$ was dissolved in 30.0 ml chloroform and added to 10.0 ml of the polymer solution prepared as described above. The solutions were stirred for 1 hr. This procedure was followed for all of the solutions resulting from syntheses with the values of t_1 of 40, 60, 90, 150, 600, and 990 minutes. Analogous polyurethane ionomer solutions were prepared with Tb, for fluorescence measurements, and with Er(III), Dy(III) and Er(III)/Yb(III)(1:1) for preparing films for near IR spectra.

Measurements.

The FTIR measurements were performed on Perkin-Elmer 1600 series FTIR. A drop of the solution was placed on KBr; and after most of the solvent evaporated, a strong nitrogen flow was applied for several minutes to drive out the remaining solvent. The florescence measurements were made on solutions of the

38

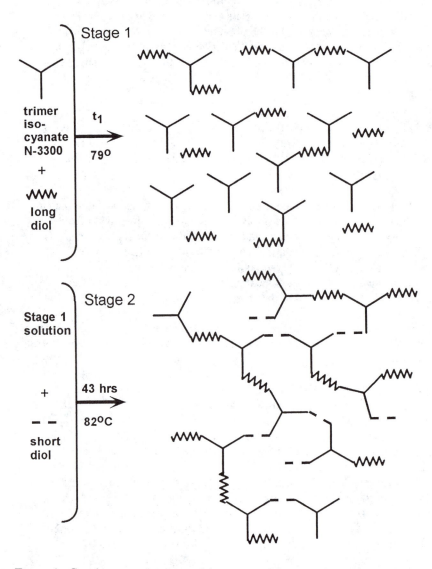

Figure 1. Graphic representation of the stages of the procedure for synthesizing acid-form polyurethanes of constant stoichiometry but varied initial prepolymer formation time.

terbium ionomer of an acid-form polyurethane (labeled PU47) and of terbium acetylacetonate in the solvent system used for the measurements of optical activity of the erbium samples. Near infrared measurements were made on films of Er(III), Dy(III), and Er(III)/Yb(III) (1:1) ionomers formed by casting solutions of the ionomers prepared with the same acid-form polyurethane. The solvent was removed at 25°C under a N_2 flow. The PU47 has a stoichiometry based on the combination of 0.890g N-3300, 0.226g L-tartaric acid and 1.219g polybutadiene diol with t_1 of 90 minutes.

Optical rotation was measured on Perkin-Elmer 241 polarimeter at 589, 578, 546, 436, 365, and 313nm. The measurement temperature was 25.5±0.5 °C, and the cell was 10.0 cm long. All solutions studied were ultracentrifuged before measurement, although they appeared clear before centrifugation.

To prepare reference solutions, 1.1190g D (or L)-tartaric acid were dissolved in 80.0 ml N,N-dimethylacetamide. Then the solutions were diluted to 250.0 ml with ethyl acetate in a volumetric flask. The resulting solutions were $4.476*10^{-3}$ g/ml. The optical rotations of the above solutions and all the polymer solutions were measured by using 1:3 N,N-dimethylacetamide/ethyl acetate mixture as reference.

In addition, 10.0ml D (or L)-tartaric acid organic solvent solutions described above were diluted by 30.0 ml chloroform. The resulting solution has D (or L)-tartaric acid concentration $1.119*10^{-3}$ g/ml. The optical rotation of the above diluted D (or L)-tartaric acid in organic solvent solutions and all the Er^{3+} ionomer solutions and all the comparison polymer solutions were measured by using 1:3:12 N,N-dimethylacetamide/ethyl acetate/ chloroform mixture as reference.

The specific rotation, $[\alpha]_\lambda$, of an optically active material is measured as the degree of rotation of polarized light per amount of material. The molecular rotation, $[m]_\lambda$, specifies this as the degree of rotation on a molar basis and is used to express the contribution of chiral constituents to the optical activity of polymeric molecules. The optical rotation of the solutions and solvents were measured at 25°C. The rotation data presented in Figs. 2 - 5 are given in terms of $[m]_\lambda$ in units of 10^2degree[decimeter(mol/ml)]$^{-1}$. The mol value is that of the tartaric acid content. Data are presented only for the cases in which D-TA or L-TA was used, because the analogous polymers based on D,L-TA, or meso-TA were not optically active. Absorption spectra and CD spectra showed no peaks in this wavelength range.

Results

The reactions were carried out with the compositions given in the previous section. For the products whose optical activity data are reported, the ratio of isocyanate functional groups to hydroxyl groups from polybutadiene diol (PBD) and tartaric acid (TA) is 1.02 : 0.404 : 0.596. This means that about 40% of the diol molecules are PBD and about 60% are TA.

The infrared spectra taken after the synthetic reactions showed that they were complete after the second stage as shown by the disappearance of the NCO band at 2274 cm^{-1}. The FTIR spectra of dried film samples of the erbium-exchanged materials showed ion-exchange and they did not show evidence for residual acetylacetone or acetylacetonate moieties.

The products whose optical activities were studied were clear liquids consisting of the polymer and the solvents noted. They are referred to as solutions, since they are clear and do not form precipitates when prepared or after ultracentrifugation under normal laboratory conditions. However, branched molecules can be highly dispersed macromolecules with closely matched refractive indices as well as dissolved species, and it is important to note this operational definition of solution. In this regard, it also can be noted that preliminary results from related work indicate that molecules with molecular weights on the order of 10^5 Daltons contribute to light scattering in similar systems (7).

The optical activities of the materials formed using L-tartaric acid are shown in Figure 2. The curves are for the acid-form polyurethanes with the indicated value of t_1, the first stage reaction time. Labels of the form L-t_1 are given for these curves. Also in Figure 2 are the optical activities of L-tartaric acid in two other solutions; water and the solvent system and concentration used for the polymers. All of the polymer solutions have exactly the same stoichiometry (ratios of constituents). It is evident that the rotation per tartaric acid unit varies strongly with the time t_1 of the first step of the synthesis. Thus, [m] can be tuned from about 40 to 110 at 365 nm by varying t_1 from 40 to 990 min. Moreover, comparison of the polymer curves with the curves for L-tartaric acid show that the rotation per TA unit is quite different when TA is incorporated in the polymer than when it is in solution alone. The difference between the curves for L-tartaric acid also confirms that the activity is solvent dependent.

The analogous results for the materials made with D-tartaric acid are shown in Figure 3. Again, it is clear that the time of the first stage of the reaction has a significant effect on the optical activity of these stoichiometrically identical compounds. It is interesting to note, as will be discussed below, that the ones with the smallest values of t_1 have optical activities that are most different from those of the dissolved tartaric acid. Further, the greater t_1 is the more similar the product's optical properties are to those of tartaric acid.

The optical activities of these polymers in more dilute solution are shown in Figure 4. These solutions were prepared to serve as the reference for measurements of the erbium ionomers. This reference is required because the ionomers were prepared and measured in an augmented solvent system at greater dilution and the optical activity of the acid-form materials is dependent on the solvent and concentration. The optical activities of the erbium ionomer solutions are shown in Figure 5.

The curves in Figure 5 are quite different from those in Figure 4. It is clear that complexation by erbium ions has changed the polymers significantly. To a first approximation, all of the materials based on a given form of tartaric acid have quite similar properties after complexation even though they were very different in the acidic form. While they are not identical sets, they are remarkably more similar after formation of the erbium ionomers.

The incorporation of the lanthanides in the ionomers clearly has an effect on the optical activity, and it is of interest to explore some of the other properties in

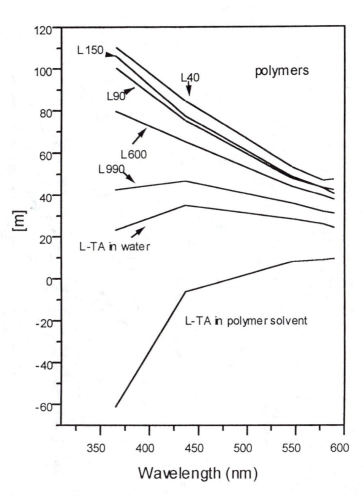

Figure 2. Optical activity of acid-form polyurethanes based on L-tartaric acid, L-tartaric acid in water, and L-tartaric acid in the polymer solvent (see text) plotted in terms of [m]$_\lambda$ in units of 10^2 degree[decimeter(mol/ml)]$^{-1}$. The mol value is that of the tartaric acid content.

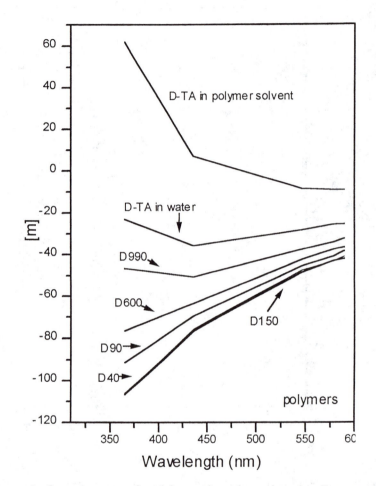

Figure 3. Optical activity of acid-form polyurethanes based on D-tartaric acid, D-tartaric acid in water, and D-tartaric acid in the polymer solvent (see text) plotted in terms of $[m]_\lambda$ in units of 10^2degree[decimeter(mol/ml)]$^{-1}$. The mol value is that of the tartaric acid content.

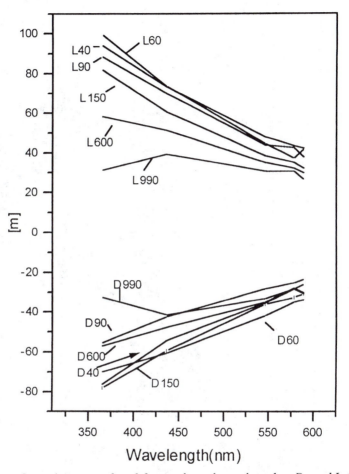

Figure 4. Optical activity of acid-form polyurethanes based on D- and L-tartaric acid in the diluted polymer solvent employed for study of their erbium ionomers(see text), plotted in terms of [m]$_\lambda$ in units of 10^2 degree[decimeter(mol/ml)]$^{-1}$.

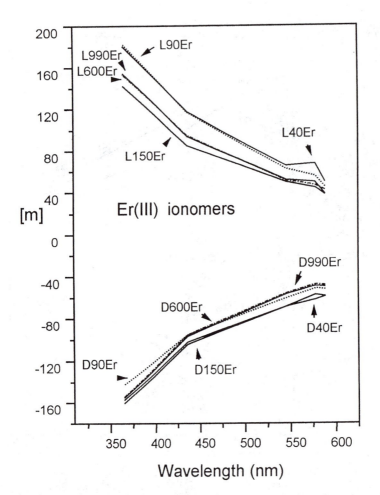

Figure 5. Optical activity of erbium polyurethane ionomers based on D- and L-tartaric acid in the ionomer solvent (see text) plotted in terms of [m]$_\lambda$ in units of 10^2degree[decimeter(mol/ml)]$^{-1}$.

a preliminary manner. The terbium form of a polyurethane acid-form polymer was prepared and the fluorescence spectrum of its solution was measured with 350 nm excitation. The spectrum of a solution of terbium acetylacetonate also was measured under the same conditions. These spectra are presented together in Figure 6. The spectra are approximately normalized in the 530 to 570 nm region of the emission spectra. This allows one to observe that the relative intensity of the bands near 540 and 585 nm is different in the two cases. The relative intensities apparently are different in the cases involving the band near 490 nm as well, but the change in the background makes that a bit harder to discern without further analysis. The elevated background shows that the polymer contributes to short wavelength scattering, but that it is not a large effect relative to the penetration of the excitation or the intensity of the fluorescence at the measured concentration.

The near infrared absorption spectra of Er(III) and Dy(III) ionomers, together with the one containing equal amounts of Er(III) and Yb(III), are shown in Figure 7. The well known Er(III) band near 1.53 μm is evident in the Er(III) and the Er(III)/Yb(III) ionomer spectra, but not in the spectrum of the Dy(III) ionomer. Indeed, there is relatively low absorption from the polymeric part of the ionomer or its overtones in the 1.5 – 1.6 μm region. The spectra in the 0.8 – 1.0 μm region exhibit the expected transitions for the ions (8), while the bands in the 1.6 – 1.8 μm region are due to overtones of the polymer's fundamental vibrations.

Discussion

An approach to interpreting the dependence of the optical activity of these novel acid-form polyurethanes on synthetic conditions is based on the extent of the reaction of the isocyanate with the polybutadiene diol during the first stage reflux. The optical properties of the products prepared with small values of t_1, in the 40 to 150 minute range, do not vary greatly. For short times, small molecules and short linear arrangements of prepolymers are expected to dominate the Stage 1 solutions. This is illustrated schematically in the top part of Figure 8. These structures apparently have reactivities which lead to incorporation of the tartaric units with a particular steric preference. However, when t_1 is much longer, the molecular structures apparently favor the incorporation of tartaric acid units in a different manner, and this affects the optical activity. One way this can happen is illustrated in the bottom part of Figure 7. This scheme is based on the following argument.

Theoretical study of stepwise polymer growth when A_3 reacts with B_2 (9) shows that the average M_W remains low through its initial stages, but, after the reaction has proceeded to a certain extent, M_W increases dramatically. In this case, take A_3 to be the t-HDI ring and B_2 to be the PBD reacted with it. Then, when t_1 is less than about 150 minutes the average M_W of the prepolymer is relatively low. Small segments dominate and they have a relatively high fraction of rings to which two TA units can add. In this case, the prepolymer would have a relatively high probability that two TA units would be added to the same ring, and the products would be the ones containing the form of tartaric acid that is favored for incorporation in that structure.

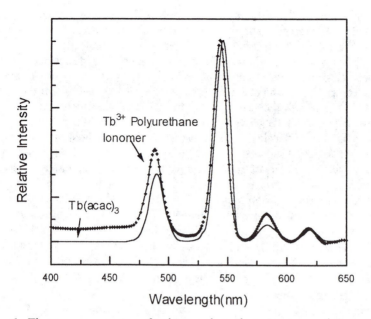

Figure 6. Fluorescence spectra of terbium polyurethane ionomer solution and terbium acetylacetonate in the ionomer solvent (see text) excited at 350 nm. Approximate normalization set in the 530 – 600 nm range.

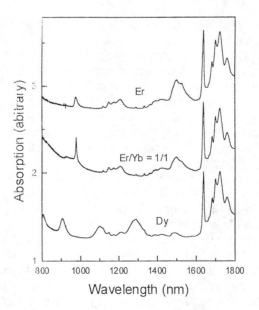

Figure 7. Near infrared spectra of films of Er(III) polyurethane ionomer (Er), Dy(III)polyurethaneionomer (Dy), and Er(III)/Yb(III) poyurethane ionomer (Er/Yb=1/1). Absorption spectra are normalized approximately to the 1.6μm ionomer band and offset for display.

Stage 1 with t₁ small (representative structures)

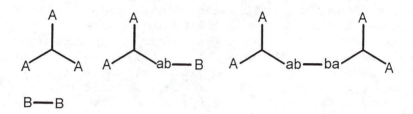

Stage 1 with t₁ large (limiting case substructure)

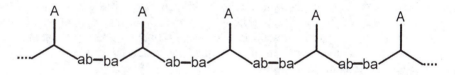

Figure 8. Schematic representation of the structural postulate concerning the distribution of isocyanate sites available for tartaric acid addition as a function of the time of the first stage of the reaction.

When t_1 is greater than about 150 minutes, larger prepolymer segments form. Since t-HDI is in great excess relative to PBD during step 1, it is expected that these prepolymer segments will be predominantly linear. One of the A sites in A_3 will remain unreacted while the other two are attached to B_2 units. That has the consequence of separating the isocyanate sites at which the TA units can add. When that separation is great enough, there is only a relatively low steric barrier to the incorporation of any form of TA. Under that circumstance, the most abundant steric form of TA is incorporated. Thus, as t_1 increases, the $[m]_\lambda$ curves for the polymers increasingly approach the curve for TA in the same solvent.

This interpretation is that different steric forms of tartaric acid can be incorporated in the polymers and that incorporation selectivity is exercised by the reactivities of the structures formed in the first stage of the reaction. The result of this selectivity is that some of the tartaric acid moieties are incorporated in energetically disfavored conformations. That is more important when t_1 is small. This disfavored conformation also appears to be the one favored by complexation of the polymer with erbium. Indeed, the optical activity of the erbium ionomers are even more unlike those of the tartaric acid itself in solution than are those of the acid-form polymers prepared with small values of t_1. Simple steric arguments suggest that the tartaric acid moiety structure with the carboxylate groups in the gauche configuration is the one which can bind with erbium in a bidentate fashion. That mode of complexation is the one that is least likely to form intermolecular crosslinks and cause precipitation. It also would be the one that would stabilize the tartaric acid moieties in what would otherwise be disfavored conformations. These interpretations are consistent with all of the observations. Further work on this system is underway to refine the interpretation and to develop these materials for optoelectronic applications.

References

1. Goodman, M.; Chen, S. C. Macromolecules **1970**, 4, 398.
2. Muller, M.; Zentel, R.; Macromolecules **1996**, 29, 1609.
3. Gu, H.; Nakamura, Y.; Sato, T.; Teramoto, A.; Green, M. M.; Jha, S. K.; Andreola, C.; Reidy, M. P. Macromolecules **1998**, 31, 6362.
4. Akagi, K.; Piao, G.; Kaneko, S.; Sakamaki, K.; Shirakawa, H.; Kyotani, M. *Science* **1998**, 282, 1683 .
5. Goodman, M.; Chen, S. C. Macromolecules **1971**, 4, 625.
6. *Advances in Urethane Ionomers*; Xiao, H. X.; Frisch, K. C., Eds.; Technomic Publ. Co.: Lancaster, PA, 1995.
7. Gu, Q.; Bartles, C.; Risen, W. M., Jr., unpublished.
8. Rajagopalan, P.; Tsatsas, A. T.; Risen, Jr., W. M. J. Polym. Sci. B. **1996**, 34, 151 .
9. Miller, D. C.; Macosko, C. W. Macromolecules **1978**, 11, 656.

Chapter 4

Perfluorocyclobutane Polymers for Optical Fiber and Dielectric Waveguides

Dennis W. Smith, Jr.[1], Adrienne B. Hoeglund[1], Hiren V. Shah[1], Michael J. Radler[2], and Charles A. Langhoff [2]

[1]Department of Chemistry, Clemson University, 219 H. L. Hunter Chemistry Laboratory, Clemson, SC 29634
[2]Corporate Research, The Dow Chemical Company, 1712 Building, Midland, MI 48674

The thermal cyclopolymerization of trifunctional and bifunctional aryl trifluorovinyl ether monomers to perfluorocyclobutane polymers affords high temperature, variable refractive index, low dielectric constant and low transmission loss materials for potential use in optical communication and microelectronic devices. Copolymerization reactions were studied to determine monomer reactivity differences and reveal the dependence of optical properties on copolymer structure. Random copolymerization affords tunable thermal and optical properties as a function of copolymer composition.

Organic polymers for applications such as dielectric waveguides and optical fiber are increasingly attractive alternatives to inorganic components in telecommunication devices.[1-5] Polymers offer flexibility, low cost fabrication and connection, high transparency in the visible and near-infrared spectra, and versatility in structure, properties, and grades for task specific integration such as local-area-network applications. However, glass-derived components remain formidable incumbents due to their lower density, high heat resistance, lower attenuation, wider bandwidth, and higher transmission speeds over the majority of common optical polymers. For example, PMMA exhibits a maximum transmission wavelength at 650 nm due to C-H vibrational overtone absorptions.[5] Most telecommunication systems require low transmission losses (<0.3 dB/cm) from 1535-1565 nm.

Due to their many complementary properties, fluoropolymers represent viable alternatives to current optical materials.[5-9] Halogenated polymers in general show

---

[†] This paper is dedicated to the memory of Charlie Langhoff.

negligible transmission losses in the range desired and fluoropolymers represent the lowest loss examples of organic polymers to date.[10] DuPont's Teflon-AF™ (T_g = 160 or 240 °C)[6] and Asahi's CYTOP™ (T_g = 108 °C)[7] are two such amorphous perfluoroplastics for which much interest in low loss optics exists. However, commercial perfluoropolymers in general do not exhibit the thermal and thermomechanical stability required for many commercial processes (e.g., >250 °C), and long term use environments (e.g., 85 °C / 85 % RH). Partially fluorinated polymers such as polyimides,[11] and fluoroacrylate networks[12] have also received much attention.

Our approach has been to combine thermally robust aromatic ether units with perfluorocyclobutane (PFCB) linkages and a unique crosslinking mechanism to give high T_g amorphous semi-fluorinated networks with tunable optical and thermal properties. PFCB polymers are prepared by the free-radical mediated thermal cyclodimerization aryl trifluorovinyl ether (TFVE) monomers from which a variety of low dielectric and thermally stable (350 °C) thermoplastic and thermosetting materials have been obtained (Scheme 1).[13-23]

Scheme 1.

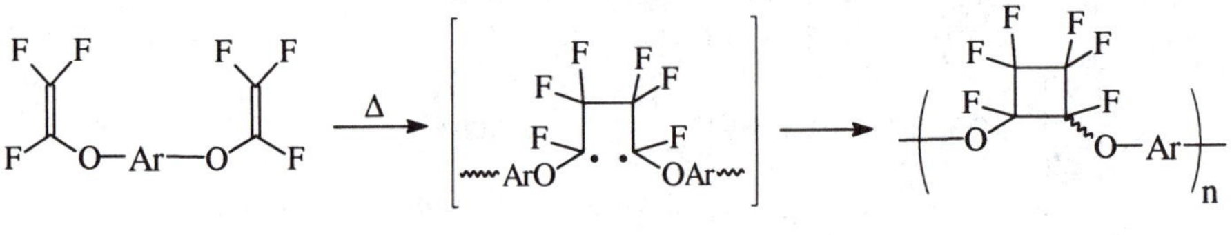

As a unique class of partially fluorinated polymers, PFCB polyaryl ethers combine the processability and durability of engineering thermoplastics with the optical, electrical, thermal, and chemical resistant properties of traditional fluoroplastics. The cyclodimerization of trifluorovinyl ethers does not require catalysts or initiators, yet proceeds thermally due to an increased double-bond strain, a lower C=C π-bond energy, and the strength of the resulting fluorinated C-C single adducts bond.[24] Step growth cycloaddition generates the more stable diradical intermediate followed by rapid ring closure giving an equal mixture of cis- and trans-1,2-disubstituted perfluorocyclobutyl linkages.[13,15] In addition, cyclopolymerization results in well-defined telechelic polymers containing known trifluorovinyl ether terminal groups.[15] PFCB polymers are easily processed from the melt or solution (e.g., spin coated) and afford high molecular weight amorphous thermoplastics or thermosets which exhibit high glass transition temperatures (T_g), good thermal stability, optical clarity, and isotropic dielectric constants around 2.3 at 10 MHz.

Trifluorovinyl ether (TFVE) monomers are traditionally prepared in three steps from commercially available phenolic precursors (Scheme 2) such as tris(hydroxyphenyl)ethane (1),[13,16] biphenol (2),[13] bishydroxyphenylfluorene (5),[16] bishydroxy-α-methylstilbene (4)[20] and very recently (hexafluoroisopropylidene)-diphenol (3).[23] Deprotonation followed by fluoroalkylation with 1,2-

dibromotetrafluoroethane in dimethylsulfoxide (< 35 °C) gives bromotetrafluoroethyl ether intermediates. Elimination of BrF with zinc metal in refluxing anhydrous acetonitrile provides trifluorovinyl aromatic ether monomers in good overall yield. Alternative strategies have also been developed which focus on the direct delivery of the trifluorovinyl aryl ether group in tact to a variety of inorganic substrates containing siloxane,[18,22] phosphine oxide,[17] and phosphazene.[21] For our optical application studies, only hydrocarbon monomers (1-3) were used. Scheme 2 illustrates the structural versatility of trifluorovinyl ether monomers prepared from phenolic precursors.

Scheme 2.

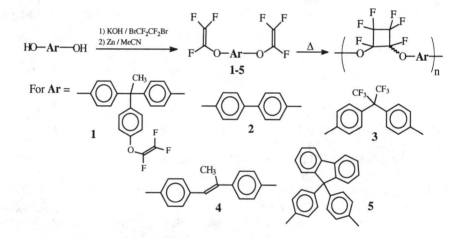

The most studied trifluorovinyl ether monomer to date has been that derived from tris(hyroxyphenyl)ethane (**1**) and – although thermosetting – exemplifies the general polymerization and excellent processability characteristics typically found for PFCB polymers. For example, monomer **1** can be solution or melt polymerized at 150 °C to a precisely controlled viscosity, molecular weight, and polydispersity.[13] The pre-network branched oligomer solutions from **1** can be spin-coated and cured giving optically clear films which exhibit exceptional planarization and gap fill for microelectronics, flat panel display, and most recently, optical waveguide applications.[25-30]

Structures can also be molded from the melt and mechanical properties such as tensile and flexural moduli for **poly1** thermoset (both near 2300 MPa) have been reported.[16] Polymerization kinetics for the polymerization of **1** has also been recently studied in detail by Raman spectroscopy.[19] Second order rate constants determined for the neat disappearance of fluoroolefins, gave half-lives which vary from 450 minutes at 130 °C to less than 10 minutes at 210 °C and activation energies of 25 kcal/mol for the cyclopolymerization.

The thermal and thermal oxidative stability of traditional PFCB thermosets, as well as the degradation mechanism and zeroth order kinetic analysis, has been reported in detail.[14] The fully cured polymer from 1 underwent catastrophic degradation at T_{onset} = 475 °C (TGA 10 °C/min in nitrogen) and gave isothermal weight loss rates of < 0.05 %/h in nitrogen and 0.7 %/h in air at 350 °C. In general, PFCB polymers are stable under inert atmosphere at 350 °C and exhibit < 1% / h weight loss. Thermal-oxidative stability is also very good for selected structures and weight loss rates can typically approach values measured under an inert atmosphere, the details of which will be reported separately.

Although the versatility of monomer structure has been greatly expanded in recent years, very little work has focused on the versatility of PFCB copolymerization.[16] Here we describe the first detailed work toward PFCB copolymers and their potential use in optical applications. The reactivity differences, relative kinetics, and copolymer thermal and optical properties as a function of copolymer composition are presented.

Experimental

Bisphenols and 1,2-difluorotetrafluoroethane were generously supplied by The Dow Chemical Company. Monomers 1 and 2 were prepared from the corresponding phenolic precursor in two steps as described previously.[13,20] Monomers 3-5 were prepared similarly from bisphenols and have been published, in part, elsewhere.[16,20,23] ^{19}F NMR was performed on a Bruker AC-200 NMR at 188 MHz. For NMR evaluations, pure monomer or monomer mixtures of equal weight were dissolved in 50 wt% mesitylene. Equal aliquots were then transferred to several nitrogen purged 5 mm NMR tubes. The tubes were then purged in N_2 and placed in a preheated oil bath at 150 °C and removed over time for NMR analysis. The reaction was quenched by cooling in room temperature water. A sealed capillary of dioxane-d_8 was added as the lock solvent and the spectral shifts were internally referenced to trichlorofluoromethane (F11). Percent conversion and average degree of polymerization (n) were estimated from ^{19}F NMR spectra using the following equations.

$$\% \text{ olefin conversion} = \frac{F_{cb}}{F_{total}} \text{ x } 100$$

Where, F_{cb} = area of PFCB signals, and F_{total} = PFCB + sum of all vinyl ($F_v = F_a + F_b = F_c$) flourine signals. The degree of polymerization (n), expressed for both linear (average functionality of 2) and branched (f > 2) polymer is then given by:

$$n_{linear} = \frac{F_{cb}}{3F_v} \quad \text{and} \quad n_{branched} = \frac{2F_{cb}}{6F_v - F_{cb}}$$

Gel Permeation Chromatography (GPC) data was collected on a Waters 2690 Alliance system equipped with a 2410 refractive index and a 996 photodiode array detector. Mixed-D and mixed-E columns (5 μm) from Polymer Laboratories were used and molecular weights were determined relative to ten narrow polystyrene standards.

Differential Scanning Calorimetry (DSC) and Thermogravimetric Analysis (TGA) were pereformed using Mettler Toledo DSC820 and TGA/SDTA851, respectively. Thermal stability of the oligomerized samples were compared by ramping the temperature to 1000 °C at 2 °C/min in a N_2 atmosphere.

For optical evaluations, polymers were prepared by heating 0.5 g of pure monomer or melt mixing 0.25 g of each under nitrogen in a glass vial on a hot plate < 100 °C. Samples were vacuum degassed in the melt and polymerizations were carried out in a N_2 purged vacuum oven maintained at 200 °C for 12-14 hrs. Upon cooling, the glass vials were broken to retrieve the samples as transparent disks. NIR/Vis spectroscopic measurements were performed on the solid polymer disks (1.3 cm dia. x 0.25cm thick) using a Shimadzu UV-3101 spectrophotometer in transmission mode. Refractive indices were measured on small polished polymer slabs (1.5 x 2.5 x 0.1 cm) with an Abbe refractometer modified to permit wavelength tuning by using a tunable monochromatic light source to illuminate the sample. Dielectric data (parallel capacitance) reported in Table I have been described elsewhere.[13]

Results and Discussion

Our goal was to exploit the well-defined polymerization mechanism offered by PFCB chemistry and prepare copolymers with tunable thermal and optical properties. The classical step growth kinetics by which PFCB polymers are formed allows for easy control of parameters important to fiber optic technology in general.[1-2] Monomer can be polymerized neat or in solution at 150 °C to precisely controlled viscosity, molecular weight, and polydispersity. Linear thermoplastic solutions or pre-thermosetting cases composed of hyper-branched oligomers can be melt processed or spin-coated. Final cure is then achieved by post baking under nitrogen or air at temperatures ranging from 235-325 °C for several hours depending on the device.

Copolymerization Kinetics

Tris(trifluorovinyloxyphenyl) ether (1) and bis(trifluorovinyloxy) biphenyl ether (2) were dissolved in a 1:1 (wt) ratio in mesitylene (50 w/w %) and thermal cyclo-copolymerization was carried out in N_2 purged sealed NMR tubes at 150 °C for 40 hours. The tubes were removed periodically and quenched to room temperature. Figure 1 shows the ^{19}F NMR spectra for a pre-gelled copolymer sample (poly1-co-2) after solution polymerization in mesitylene for 24 h. Figure 2 graphs the fluoroolefin conversion by ^{19}F NMR vs. time for the copolymer and similarly prepared homopolymers from 1 and 2.

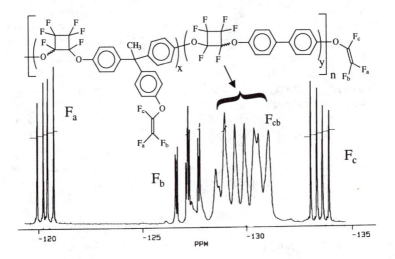

Figure 1. ^{19}F NMR spectra of **poly1-co-2** (1:1 wt) after 24 h at 150 °C (50 wt% in mesitylene, calculated n = 17, M_n = 8000) and idealized linear structure of the pre-gelled branched oligomer.

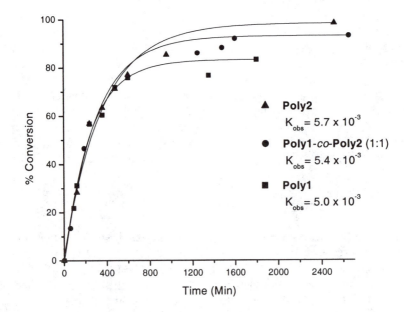

Figure 2. Fluoroolefin conversion vs. time for solution (50 wt % mesitylene) polymerization at 150 °C by ^{19}F.

Endgroup analysis by ^{19}F NMR has been a reliable asset in PFCB chemistry (Figure 1).[15] Although the three sets of nonequivalent vinyl fluorine signals (dd near − 120, -127, -134 ppm) overlap surprisingly close, it was possible to determine from this data that the relative rate of disappearance for the two monomers is nearly identical. Integration of the multiple nonequivalent cyclobutane fluorines vs. the clearly resolved vinyl fluorine, F_a or F_c, gave a n = 17 or M_n = 8000 for the ideal linear structure shown. In contrast, M_n = 3200 and M_w/M_n = 5 was measured by GPC (vs. PS) which indicates substantial branching has occurred.

From the data in Figure 2, the rates of homopolymerization for monomers **1, 2**, and their 1:1 (wt) copolymerization (**poly1-co-2**) are essentially identical beyond one half life at 150 °C where the observed second order rate constants ranged from 5.0×10^{-3} to 5.7×10^{-3} min^{-1} by ^{19}F NMR. This data agrees well with previous data determined for the polymerization of **1** by Raman spectroscopy.[19] The divergence of the polymerization rate above 60 % conversion is due to gellation of compositions containing trifunctional monomer **1**. Trifluorovinyl ether compositions with functionality greater than two are predicted to behave classically as was found for **1** which gels at 50 % conversion (1/f-1).[19] Likewise, the polymerization of **1-co-2** should gel near 67 % functional group conversion.

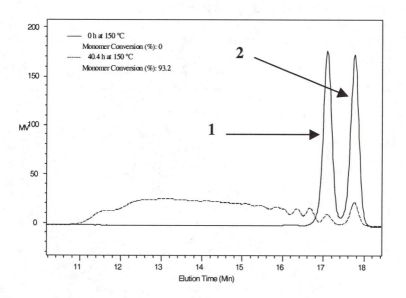

Figure 3. GPC profiles of starting monomer mixture and **poly1-co-2** *(1:1 wt) after 40 h at 150 °C.*

In addition to functional group conversion determined by ^{19}F NMR, monomer conversion was also calculated from GPC profiles as shown in Figure 3. The rate of monomer consumption for **1-co-2** (1:1) copolymerization after 40 h at 150 °C (93 % olefin conversion) appears essentially equal. This data, along with NMR endgroup analysis and single glass transition temperatures (*vide infra*) provided good evidence that **poly1**-*co*-**2** is random and that variable copolymer compositions can be easily prepared by simple choice of initial comonomer ratio.

Thermal Characterization

Glass transition temperatures were determined by DSC and are listed in Table I. Copolymers prepared from trifluorovinyl aryl ether monomers **1**-**2** gave single glass transitions at values predicted by the Fox equation.[31] For example, **poly1**-*co*-**2** (1:1) gave $T_g = 220$ °C, whereas the theoretical T_g based on homopolymer values is 224 °C. The copolymer from monomers **1** and **3** has also been prepared yet the exact composition of the copolymer is unknown due to, presumably, the added volatility of monomer **3** containing twice the number of fluorine atoms. A $T_g = 174$ °C was measured for **poly1**-*co*-**3** (1:1 wt charge) which is slightly lower than the predicted value. Likewise the refractive index of the copolymer was higher than the theoretical value (*vide infra*) which is consistent with loss of monomer **3** during copolymerization.

Thermal stability was accessed by TGA in a dynamic mode initially. In order to ensure that all samples were void of mesitylene solvent, polymers were heated at 200 °C for three hours and then cooled to 35 °C. The samples were then heated to 1000 °C at 2 °C/min in a nitrogen purged environment. From the TGA data (Figure 4), the thermal stability of the homopolymers and co-polymer are essentially the same under these conditions (T_{onset} ca. 450 °C). Although the degradation onset temperature of **poly2** is slightly lower than that for **poly1**, the copolymer shows no significant difference in the rate of degradation.

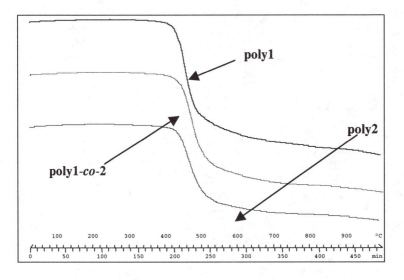

Figure 4. TGA analysis of selected PFCB polymers in nitrogen at 2 °C/min.

Optical Properties

Figure 5 shows the optical attenuation of the homopolymers and **poly1-*co*-2** in the visible and NIR spectral regions. The observed optical loss profile of the copolymer is intermediate to that of the two homopolymers, as expected. In addition, the PFCB homopolymers and the copolymer exhibit low loss in the wavelength range suited for optical waveguide applications (1535-1565 nm). Although quantitative loss measurements have not been determined for **poly2** and **poly1-*co*-2**, an attenuation of 0.25 dB/cm at 1515-1565 nm was reported previously for the homopolymer from **1**.[27] The relative loss data presented here predicts that a similar value may be obtained for the copolymer and a slightly higher loss for **poly2**.

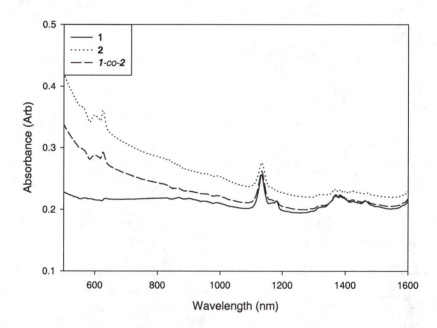

Figure 5. Transmission loss data selected PFCB polymers by NIR / Vis spectoscopy.

The dependence of refractive index on wavelength has also been measured for homopolymers from monomers **1-3** and **poly1-*co*-2** (Figure 6). As mentioned earlier, copolymers from **1** and **3** have also been prepared yet are not included here due to their uknown exact composition. The simple rule of mixtures is maintained for these initial PFCB copolymers and it appears that an accurate range of variable RI materials can be achieved by simple choice of comonomer composition. By varying these amounts, it is possible to obtain tunable polymers which allows for a greater range of use. Table I further summarizes some selected properties.

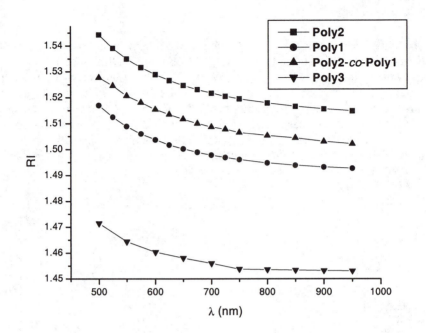

Figure 6. Refractive Index vs. wavelength for selected PFCB polymers.

Table I. Selected Properties of PFCB Polymers

PFCB Polymer from	M_n	M_w/M_n (GPC)	T_g (°C)	n (900 nm)[e]	ε (10 kHz)
1	-	-	350[c]	1.494	2.35
2	50,000[a]	2	165[c]	1.517	2.41
3	60,000	1.9	120	1.452	-
1-co-2 (1:1) (pregel 24h rxn)	8,000[a] / 3200[b]	5	220[d]	1.503	-

[a]NMR end group analysis. [b]GPC vs. PS. [c]DMS. [d]Cured to 350 °C by DSC. [e]Abbe refractometer (see experimental).

Microfabrication via Soft-Lithography

Recently, our efforts have focussed on development of non-photolithographic processing methods for fabrication of micropatterns suitable for optical waveguides and diffractive elements. Microfabrication using non-photolithographic techniques (*soft lithography*) has been pioneered by Whitesides et. al.,[32] and recently used in our labs with polynaphthalene networks.[33] These techniques involve preparation of poly(dimethylsiloxane) (PDMS) molds and/or stamps by casting PDMS on a photolithographically generated silicon master with the desired micropattern. The PDMS mold can then be used in a variety of ways to replicate the micropatterns on a given substrate. Two such examples, namely microcontact printing (μCP) and micromolding in capillaries (MIMIC), are shown schematically in Figure 7 (A).

Both soft lithography techniques were found to work very well with PFCB polymers (Figure 7A). However, during our investigations we found that the PDMS negative-mold can be eliminated for PFCB polymers. Instead PFCB replicas can be generated by directly molding the polymer against silicon masters as shown schematically in Figure 7 (B). Since PFCB polymers provide low interfacial surface energy, the replicated structures can be easily lifted off from the master without irreversible adhesion or defects. Homopolymers or copolymers can be melt or solution cast on to the silicon masters containing 0.5 μm features. However, the lift-off process is simplified to a large extent by solution-casting due to the plasticizing effect of the solvent which can be subsequently removed by vacuum drying.

Although microfabrication procedures for the PFCB homopolymer 1 has been recently reported for large-core waveguides (5-50 μm),[30] the "negative mold-free" technique highlighted here demonstrates feature reproduction at submicron scales for the first time. This technique also yields free standing structures with variable RI suitable for applications such as adaptive optics.[34] Efforts are currently underway to prepare PFCB based refractive/diffractive optical and other hybrid devices and measure their performance.

Conclusions

The thermal cyclopolymerization of aryl trifluorovinyl ether monomers to perfluorocyclobutane polymers affords high temperature, variable refractive index, and low transmission loss materials for potential use in optical communication devices. NMR and GPC analyses indicate that the rate of homopolymerization and copolymerization are essentially identical and the resulting copolymers are completely random. Copolymers with variable refractive indices (1.3-1.5), glass transition temperatures (165-350 °C), and long-term 350 °C thermal stability were accessible by simple choice of comonomer composition. Direct sub-micron transfer molding of PFCB optical gratings without the use of PDMS molds has also been demonstrated.

Acknowledgements

We thank Clemson University, The Dow Chemical Company, 3M Corporation (3M Pretenure Award for DWS), and the NASA EPSCoR South Carolina Space Grant Consortium for supporting this work. We also thank D. Babb and R. Snelgrove of

(A)

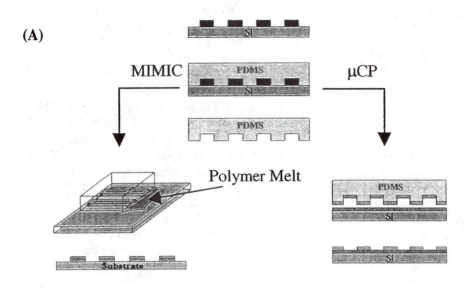

Polymer Melt

(B)

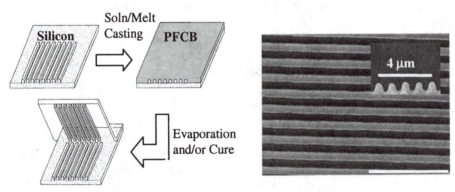

(C)

Figure 7. (A) Micromolding in capillary (MIMIC) and microcontact printing (μCP). (B) "Negative mold-free" microfabrication of PFCB films. (C) SEM Surface and cross-section (inset) of a PFCB diffraction grating (0.5 μm wide lines).

The Dow Chemical Company for technical support and expertise, Mettler-Toledo for the donation of the DSC820 and TGA/SDTA851 thermal analysis system, and D. DesMarteau (CU) for helpful advice and instrument support.

References

1. For a succinct tutorial see: Harmon, J.P. *Polym. Prepr. (Am. Chem. Soc. Div. Polym. Chem.)* **1999**, *40(2)*, 1256, and references therein.
2. Marcou, J., Ed. *Plastic Optical Fibers*; John Wiley and Sons: New York, **1997**.
3. Koeppen, C.; Shi, R.F. Chen, W.D.; Garito, A.F. *J. Opt. Soc. Am. B*, **1998**, *15(2)*, 727.
4. Levy, A.C.; Taylero, C. In *Encycl. Polym. Sci. & Eng.* J. Kroschwitz, Ed.; Vol 7., Wiley: New York, **1987**, pp. 1-15.
5. Kaino, T.; Yokohama, I.; Tomaru, S.; Amano, M.; Hikita, M. In *Photonic and Optoelectronic Polymers* Jenekhe, S.A.; Wynne, K.J. Eds., ACS Symp. Ser. **1997**, *672*, 71.
6. Resnick, P.R.; Buck, W.H. In *Modern Fluoropolymers*, Scheirs, J., Ed.; Wiley: New York, **1997**, p. 397.
7. Sugiyama, N. In *Ibidi*, p. 541.
8. Giaretta, G.; White, W.; Wegmuller, M.; Onishi, T. *IEEE Phot. Tech. Let.* **2000**, *12(3)*, 347.
9. Nihei, E.; Ishigure, T.; Tanio, N.; Koike, Y. *IEICE Trans. Electron.* **1997**, *E80-C(11)*, 117.
10. Groh, W. *Macromol. Chem.* **1988**, *189*, 1261.
11. Matsuura, T.; Ando, S.; Sasaki, S.; Yamamoto, F. *Electron. Lett.* **1993**, *29(3)*, 269.
12. Hale, A.; Quoi, K.; DiGiovanni, D. *Polym. Prepr. (Am. Chem. Soc. Div. Polym. Chem.)* **1998**, *39(2)*, 978.
13. Babb, D.A.; Ezzell, B.R.; Clement, K.S.; Richey, W.F.; Kennedy, A.P. *J. Polym. Sci.: Part A: Polym. Chem.* **1993**, *31*, 3465.
14. Kennedy, A.P.; Babb, D.A.; Bremmer, J.N.; Pasztor, Jr., A.J. *J. Polym. Sci.: Part A: Polym. Chem.*, **1995**, *33*, 1859.
15. Smith, Jr., D.W.; Babb, D.A. *Macromolecules* **1996**, *29*, 852.
16. Babb, D.; Snelgrove, V.; Smith, Jr., D.W.; Mudrich, S. In, *Step-Growth Polymers for High Performance Materials*, J. Hedrick, J. Labadie, Eds., ACS Sym. Ser., 624, American Chemical Society, Washington, D.C. **1996**, p. 431.
17. Babb, D.A.; Boone, H.; Smith, Jr., D.W.; Rudolf, P. *J. Appl. Polym. Sci.* **1998**, *69*, 2005.
18. Ji, J.; Narayan, S.; Neilson, R.; Oxley, J.; Babb, D.; Rondan, N.; Smith, Jr., D.W.; *Organometallics* **1998**, *17*, 783.
19. Cheatham, C.M.; Lee, S-N.; Laane, J.; Babb, D.A.; Smith, D.W., Jr. *Polymer International*, **1998**, *46*, 320.

20. Smith, Jr., D.W.; Boone, H.W.; Traiphol, R.; Shah, H.; Perahia, D. *Macromolecules*, **2000**, *33(4)*, 1126.
21. Neilson, R.H.; Ji, J.; Narayan-Sarathy, S.; Smith, D.W.; Babb, D.A. *Phosphorous, Sulfur and Silicon,* **1999**, *144-146*, 221.
22. Smith, Jr., D.W.; Ji, J.; Narayan-Sarathy, S.; Neilson, R.; Babb, D. In *Silicones and Silicone Modified Materials*, ACS Symp. Ser. 729, American Chemical Society, Washington, D.C. **1999**, p. 308.
23. Smith, D.W., Jr.; Shah, H.V.; Johnson, B.R. *Polym. Prepr. (Am. Chem. Soc., Div. Polym. Chem.)* **2000**, *41(1)*, 60.
24. Smart, B.E. In *Organofluorine Chemistry Principles and Commercial Applications*; Banks, R.E.; Smart, B.E.; Tatlow, J.C., Eds.; Plenum Press: New York, **1994**, p. 73.
25. Tumolillo, T.A., Jr.; Thomas, A.; Ashley, P.R. *Appl. Phys. Lett.*, **1993**, *82(24)*, 3068.
26. Townsend, P.; Shaffer, E.; Mills, M.; Blackson, J.; Radler, M., In *Low-Dielectric Constant Materials II*, A. Lagendijk, H. Treichel, K. Uram, A. Jones, Eds., *Mat. Res. Soc.* **1996**, *443*, 35.
27. Fischbeck, G.; Moosburger, R.; Kostrzewa, C.; Achen, A.; Petermann, K. *Electon. Lett.* **1997**, *33(6)*, 518.
28. Oh, M.; Lee, M.; Lee, H. *IEEE Photonics Tech. Lett.* **1999**, *11*, 1144.
29. Siebel, u.; Hauffe, R.; Petermann, K. *IEEE Photonics Tech. Lett.* **2000**, *12(1)*, 40.
30. Lee, B.; Kwon, M.; Yoon, J.; Shin, S. *IEEE Photonics Tech. Lett.* **2000**, *12*, 62.
31. Fox, T.G. *Bull. Amer. Phys. Soc.* **1956**, *1*, 123.
32. Qin, D.; Xia, Y.; Rogers, J.; Jackman, R.; Zhao, X.; Whitesides, G.M. in *Microsystem Technology in Chemistry and Life Sciences*, vol. 194, Manz, A and Becker, H., Eds. Springer-Verlag, Berlin, **1998**.
33. Shah, H.; Brittain, S.; Huang, Q.; Hwu, S.J.; Whitesides, G.M.; Smith, D.W., Jr. *Chem. Mater.* **1999**, *11*, 2623.
34. Hubin, N.; Noethe, L. *Science* **1993**, *262*, 1390.

Chapter 5

Use of Hydroxy Functional Fluoropolymer Resins in Free Radical UV Curable Coatings

G. K. Noren

DSM Desotech Inc., 1122 St. Charles Street, Elgin, IL 60120

Blends of a hydroxy functional fluoropolymer resin with conventional UV curable acrylate diluents were prepared and cured with UV light to form semi-interpenetrating networks (SIPN). After cure the amount of fluoropolymer resin extracted from the films after 5 days with toluene at ambient temperature ranged from 10 to 50 percent of the original amount of the fluoropolymer resin and was proportional to amount of the fluoropolymer resin in the formulation. Thus, some of the fluoropolymer resin was either physically or chemically incorporated into the crosslinked network. A urethane acrylate oligomer was synthesized from a second hydroxy functional fluoropolymer resin. The oligomer gave good mechanical properties when cured and was compatible with standard acrylate diluents. The cure speed of a model formulation containing the fluoropolymer resin based urethane acrylate oligomer was measured by the modulus ratio at doses of 0.2 and 0.5 J/cm^2 and was found to be 0.9. Environmental stability testing using QUV light and 24 hours of water soaking showed no advantage due to the presence of the fluorine substituted oligomer when compared to a conventional urethane acrylate.

Introduction

Several types of hydroxy functional fluoropolymer resins (FPR) are commercially available and could offer not only flexibility but also the other beneficial properties of fluorine containing polymers. One class of commercially available hydroxy functional fluorocarbon resins are relatively highly fluorinated (50-60% of the molecule) fluoropolymers containing terminal hydroxyl groups[1-3] and are quite expensive. Another class of commercially available hydroxy functional fluorocarbon resins are prepared from chlorotrifluoroethylene, vinyl ethers and hydroxybutyl vinyl ether and contain pendent hydroxyl groups but have less fluorine (25-30% of the molecule)[4] and are lower priced. Thermal crosslinking of both types of resins with either isocyanates[2,4] or melamine-formaldehyde resins[4] yields films with superior weatherability, good chemical resistance and adhesion to most substrates. (Meth)acrylate functionalized oligomers of both types have also been reported.[5-11] In a previous paper we have reported on the investigation of the use of the fluorocarbon resins containing pendent hydroxyl groups in cationic curable formulations.[12] In this paper we continue the investigation of fluoropolymer resins by incorporating them into free radical curable formulations and the synthesis of the urethane acrylate of a fluoropolymer resin (FPR3). Our objective in this work was to produce fluorine containing free radical UV curable coating formulations that would have improved weathering and moisture resistance by using the lower cost fluoropolymer resins.

Experimental

The chemicals used in the oligomer synthesis and formulations were used as received from the suppliers. Fluoropolymer resins (Table 1) were obtained from Zeneca Resins.[7] The formulations were prepared by simply blending the ingredients in a 4 oz brown glass bottle and mixing them on a laboratory shaker at room temperature for about 2 hours. No antioxidants, hindered amines or other types of stabilizers were used.

The synthesis of the FPR3 urethane acrylate (FPR3UA) was accomplished in two steps. First, a 1 to 1 adduct of 2-hydroxyethyl acrylate (19.45g; 0.1675 mol) with isophorone diisocyanate (37.23g; 0.1675 mol) was prepared by slowly adding the alcohol to the diisocyanate at 40°C and reacted until the percent isocyanate reached a value of 12.4%. This adduct (56.7g; 0.1675 mol) was then added to 243.3g of FPR3 (0.2819 hydroxyl equivalents) at 40°C, then slowly heated to 60°C and held until the percent isocyanate was less than 0.2% which took about 2 hrs. This ratio of NCO/OH gives an acrylate functionality of about 2.4. The viscosity of the resulting oligomer was 66,000 mPa•s at about 71.6% NVM in xylene.

Samples for testing were prepared on glass plates using 75 (3.0 mil) Bird film applicator. The films were allowed to stand in a fume hood to evaporate the xylene. After about 20 minutes there was no further odor of xylene. The films were then cured by exposure to UV light from a Fusion Systems model F450 curing unit with a 120 w/cm (300 w/in) "D" lamp. This unit was mounted on a variable speed conveyor (4 to 75 m/min; 13 to 225 ft/min) and was capable of delivering a dose of 0.12 to 2.0 J/cm^2 in a single pass as measured with a UV Process Supply Compact Radiometer.

Toluene extractables were determined using 2" by 2" squares of 3 mil films which were cured under nitrogen using a Fusion "D" lamp at a dose of 1 J/cm^2. The initial weight was determined and the 2" by 2" film squares were placed in 4 oz glass jars with

110 ml of toluene (the fluoropolymer resins are soluble in toluene) at ambient temperature for 5 days. Then the films were separated from the toluene by filtration, dried and the final weight determined.

Tensile measurements were recorded using an Instron model 4201. Data was analyzed using Instron System 9 software. Tensile specimens were prepared by cutting 1.25 cm (0.5 in) wide strips of the coatings (75 μm; 3.0 mil) cured on glass plates using 2 passes at 1 J/cm^2. A 5.08 cm (2.0 in) gauge length was used with a crosshead speed of 2.54 cm (1.0") elongation per minute. The secant modulus at 2.5% elongation was recorded. A minimum of five tensile measurements were made for each sample.

Dynamic Mechanical Analysis experiments were performed on a Rheometrics Solids Analyzer RSA II at a frequency of 1.0 radian/second with measurements taken every 2° C. Films were prepared by the same method as for Instron testing and preheated at 80° C for 5 minutes to allow ease of stretching and equalize forces across the sample.

Two inch squares of the cured (2 passes at 1 J/cm^2) films (75 μm; 3 mil) were exposed to light from a QUV lamp which has most of it's radiation in the 250 to 350 nm range. Color changes were measured at 0,1, 2, 3, 6, and 10 days using the Macbeth Color-Eye Series 1500 colorimeter, using the machine calibration standard X, Y and Z values.

Water absorption was determined was determined by two methods. In "Method A", 3 inch squares of the cured 3 mil films were dried at 60°C for 1 hour, cooled in a desiccator, weighed (W_1) and placed in a jar containing 125 ml of distilled water at ambient temperature. Periodically the films were removed, patted dry and weighed. The percent water absorption data was plotted against time. After about 2 weeks the films were dried at 60°C for 1 hour, cooled in a desiccator, and the final weight (W_2) determined. The total percent weight lost was calculated by subtracting W_2 from W_1, dividing by W_1 and multiplying the result by 100. In "Method B", 2 inch squares of 3 mil films cured using 2 passes at 1 J/cm^2 were first equilibrated at 50% Relative Humidity and 23°C for 48 hours and weighed to get the initial weight (W_1). Then the films were soaked in deionized water for 24 hours, patted dry and quickly weighed (W_2). Then the films were dried in a vacuum (< 5 mm Hg) oven at 23°C for 24 hours and weighed (W_3). The percent water absorbed was calculated by subtracting W_3 from W_2, dividing by W_1 and multiplying the result by 100.

Results and Discussion

Four samples of fluoropolymer resins were obtained and the properties are shown in Table I.[4] Since a typical optical fiber coating has a refractive index of about 1.51 as a liquid and 1.53 in a cured film,[13,14] it was of interest to investigate the refractive index of the fluoropolymer resins. The refractive indices of the fluoropolymer resins were determined using the linear relationship for the refractive index of two component mixtures. Thus, the refractive index of the neat resin can be determined by measuring the refractive index of the solution and, knowing both the refractive index of the solvent (xylene) and the percent NVM of the solution the refractive index of the neat resin can be determined by linear extrapolation to 100% NVM. The refractive index determined

Table I. Properties of Fluoropolymer Resins

Fluoropolymer Resin	NVM (%)	Viscosity (mPa•s)	MW (Daltons)	Hydroxyl EW (Daltons)	Funct.	Tg (°C)
FPR1	65	19,000	6,000	623	9.6	37
FPR2	65	16,000	6,000	701	8.6	37
FPR3	65.2	4,000	3,500	863	4.1	37
FPR4	100	Solid	10,000	1,080	9.3	55

in this way are; 1.47, 1.46 and 1.46 for FPR1, FPR2 and FPR3 respectively. As would be expected these results are somewhat higher than the refractive index of poly(trifluorochloroethylene) which is 1.42-1.43.[15] A fluorine containing optical fiber coating was reported to have a refractive index of 1.39 as a liquid and 1.41 in the cured film.[13,14] The refractive indices of the fluoropolymers are too low for use on standard glass optical fiber and poly(methyl methacrylate) plastic fiber but may be useful for polystyrene or polycarbonate plastic fiber or other applications requiring unique optical properties.

Physical Blends (SIPN)

In order to investigate the use of these fluoropolymer resins into UV curable coating formulations a semi-interpenetrating polymer network (SIPN) type system was examined first. A Simplex design containing tripropyleneglycol diacrylate (TPGDA), trimethylolpropane triacrylate (TMPTA) and one of the fluoropolymer resins (FPR3) was used. Ten formulations were prepared and 5% of the photoinitiator (Darocur® 1173) was added based on the total solids of the formulation. The final compositions after solvent evaporation are shown in Table II. Films were prepared on glass plates from each formulation using a 3 mil Bird bar. The films were cured under nitrogen using a Fusion "D" lamp at a dose of 1 J/cm². The appearance results of the cured films from the design are shown in Figure 1. Six of the films containing high amounts of TMPTA or FPR3 were very hard and brittle and fractured when being removed from the glass plates. It would be of interest to substitute a monofunctional acrylate for the TMPTA to increase the flexibility of the films for further work. However, only four of these films could be tested. Sample (2" by 2" squares) of these four intact films were extracted with 110 ml of toluene (the fluoropolymer resins are soluble in toluene) at ambient temperature for 5 days and the amount of extractable material was determined. From the results, shown in Figure 2, it was found that the amount of fluoropolymer resin extracted with toluene

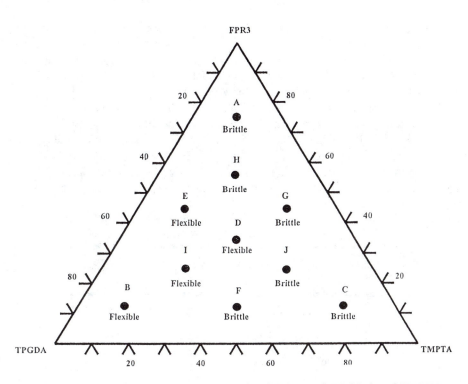

Figure 1. Film Characteristics of Blends of FPR3 with TPGDA and TMPTA.

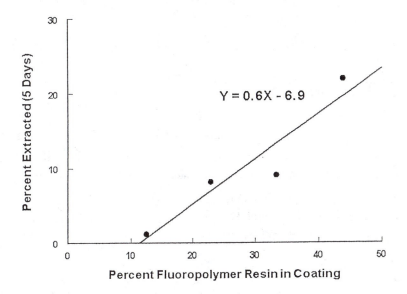

Figure 2. Toluene Extraction of Cured Films That Contain an Unmodified Fluoropolymer Resin.

was proportional to amount of fluoropolymer resin in the original formulation. The amount of fluoropolymer resin extracted ranged from 10 to 50 percent of the original amount of the fluoropolymer resin which indicated that some of the fluoropolymer resin was either physically or chemically incorporated into the crosslinked network. Due to the extractability of the fluoropolymer in this experiment it the SIPN approach would not be satisfactory for optical fiber coatings but may still be useful for less rigorous applications. In order to prepare more flexible films for the SIPN study several attempts were made to dissolve (FPR4; 100% NVM)) in mixtures of ethoxylated trimethylolpropane triacrylate (EOTMPTA) and TPGDA but even at levels as low as 20% the fluoropolymer resin would not dissolve. The fluoropolymer resin (FPR4) was also not soluble at 20% in tetrahydrofurfuryl alcohol acrylate (THFA) or phenoxyethyl acrylate (PEA).

Table II. Compositions Used in the Simplex Design Study

Formulation	FPR3[1] (%)	TPGDA (%)	TMPTA (%)	Darocur 1173[2] (%)
A	75.00	12.50	12.50	5.00
B	12.50	75.00	12.50	5.00
C	12.50	12.50	75.00	5.00
D	33.40	33.33	33.33	5.00
E	43.75	43.75	12.50	5.00
F	12.50	43.75	43.75	5.00
G	43.75	12.50	43.75	5.00
H	54.20	22.90	22.90	5.00
I	22.90	54.20	22.90	5.00
J	22.90	22.90	54.20	5.00

1. FPR3 is 65.2% NVM in xylene.
2. Darocur 1173 percentage is based on resin solids.

Acrylate Functionalized

A second objective of this work was to prepare acrylate functional fluorine containing oligomers that would have improved weathering and moisture resistance from these fluoropolymer resins. Four methods of (meth)acrylate functionalization of these fluoropolymers have been reported. They include: 1) reaction with acryloyl chloride,[7] 2) reaction with succinic anhydride followed by reaction with glycidyl methacrylate,[8] 3) reaction with 2-isocyanatoethyl methacrylate,[9] and 4) reaction with 2,6-diisocyanatocaproic acid methyl ester (LDI) followed by reaction with hydroxyethyl acrylate (HEA).[10] Based on commercial considerations we chose method 4 but substituted isophorone diisocyanate (IPDI) for LDI. A two step reaction as shown in Figure 3 was used. The HEA was reacted with IPDI to form a 1 to 1 adduct. This adduct was further reacted with the chosen fluoropolymer resin. In spite of using an isocyanate with regioselective isocyanate groups and an isocyanate to hydroxyl ratio of as low as 0.59 to reduce the possibility of gelation, both FPR1 and FPR2, which had functionalities > 8, produced gels. However, FPR3 with a functionality of about 4.1 gave a non-gelled urethane acrylate (FPR3UA) at a NCO/OH ratio of 0.59. The FPR3UA product had a viscosity of 66,000 mPa•s at about 71.6% NVM. The oligomer solution was compatible with TMPTA in a 50/50 weight by weight blend. Since this blend still contained solvent, the xylene was removed on the roto-evaporator at full vacuum and 40° C for 2 hours. The mixture was still compatible demonstrating the feasibility of producing a 100% NVM system.

In order to test the utility of the FPR3UA oligomer, the three experimental formulations shown in Table III were prepared. Coating A contains the FPR3UA oligomer solution blended with 5% based on solids of Darocur® 1173 as the photoinitiator. (Note: due to the fact that FPR3 is supplied as a solution in xylene the resultant urethane acrylate is

Table III. Model Coating Formulations

| | Coating Compositions | | |
Materials	A	B	C
FPR3UA (parts)	71.6	60.0	
Xylene (parts)	28.4	47.7	
CN 963[1] (parts)			60.0
TPGDA[2] (parts)		40.0	40.0
Darocur 1173[3] (parts)	5.0	5.0	5.0

1. A hard polyester aliphatic urethane acrylate based on a branched polyester with a MW of about 1100 to 1300 Daltons, a functionality of 2 and a Tg of 20 to 40°C.
2. Tripropylene glycol diacrylate.
3. 2-Hydroxy-2-methyl-1-phenyl-propan-1-one.

71.6% NVM in xylene) Coating B contains 60% oligomer solids of the FPR3UA, 40% TPGDA and 5% based on solids of Darocur® 1173. Coating C contains a conventional urethane acrylate oligomer as a control and consists of 60% CN 963 (Sartomer) and 40% TPGDA and uses 5% Darocur® 1173. The CN 963 is reported to be a hard polyester aliphatic urethane acrylate based on a branched polyester with a MW of about 1100 to 1300 Daltons, a functionality of 2 and a T_g of 20 to 40° C.[16]

Films for testing tensile strength, elongation, modulus and dynamic mechanical properties were prepared on glass plates using a 3 mil (75 μm) Bird bar and allowed to stand for 20 minutes to allow the evaporation of xylene from the film. The films were then cured at 2 J/cm^2 (Coating A) or 1 J/cm^2 (Coatings B and C) using a Fusion "D" lamp under nitrogen. No xylene odor could be detected after the films were cured. The testing results are shown in Table IV. The published data for a conventional optical fiber coating and fluorine containing optical fiber coating are provided for comparison.[13,14] Testing of the films from Coating A yielded a tensile strength of 12 MPa with 32% elongation and a modulus 328 MPa compared with a tensile strength of 16 MPa with an elongation of 33% and a modulus of 353 MPa for Coating B. These values are not much different from those obtained for the neat oligomer and indicate little effect on the physical properties due to the addition of TPGDA. Dynamic Mechanical Analysis of Coatings A and B also showed very little difference. Physical testing of films from Coating C gave a tensile strength of 40 MPa with an elongation of 18% and a modulus of 20 MPa. These results show the urethane acrylate from the fluoropolymer was more flexible and had a higher modulus than the conventional urethane acrylate. Although Coating B has a lower modulus and higher elongation than the conventional optical fiber coating, these properties are equal to those reported for the fluorine containing optical fiber coating.

Table IV. Properties of Model Coatings Compared with Control Coatings

Property	A	B	C	OF[1]	F-OF[2]
Cure Conditions:		Fusion D lamp, Nitrogen			
Wet Film Thickness (mil)	3	3	3	3	3
Cure Dose (J/cm^2)	2	1	1	1	UM[3]
Tensile Strength (MPa)	12	16	40	26	21
Elongation (%)	32	33	18	16	36
Modulus (MPa)	328	353	20	800	550
Equilibrium Modulus (MPa)	6.9	13.1	19.6	53	12
T_g by Tan δ_{max} (°C)	54.0	53.5	55.1	60	ND[4]

1. Typical optical fiber coating.[13,14]
2. Fluorine containing optical fiber coating. [13,14]
3. SB = Cured to 100% of it's ultimate modulus.
4. Not determined.

In order to test the cure speed of Coating B, films were again prepared on glass plates using a 3 mil Bird bar but this time cured at various dosages using the Fusion "D" lamp under nitrogen after the xylene was allowed to evaporate for 20 minutes. The modulus values were then measured at the various doses and the data plotted in Figure 4. The interesting result is the almost immediate attainment of a maximum modulus. One estimate of the cure speed of a coating is the ratio of the modulus at a low dose such as 0.2 J/cm^2 to the modulus at 1.0 J/cm^2, which is assumed to be fully cured. However, in this case, this ratio would be greater than one. Therefore, an estimate of the cure speed

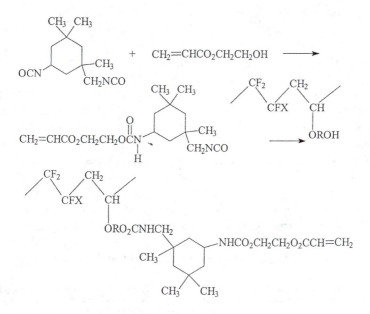

Figure 3. Synthesis of the Urethane Acrylate of a Fluoropolymer Resin.

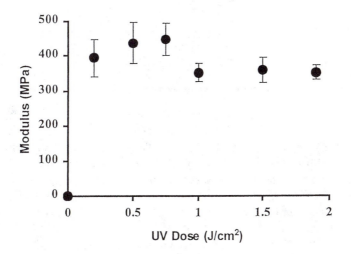

Figure 4. Dose vs Modulus Results for UV Cured FPR3UA Based Formulation.

was made by calculating the ratio of the modulus at 0.2 J/cm^2 to the modulus at 0.5 J/cm^2 which is equal to 0.9. This high rate of modulus development is probably due to the high functionality of the FPR3UA.

Three mil films of Coating B and Coating C were prepared in order to study the effect of QUV exposure. The xylene was allowed to evaporate for 20 minutes and the films were cured at 2 J/cm^2 (D lamp) under nitrogen and then exposed to the QUV lamps. Color changes (yellowing) were monitored over a 13 day period using a MacBeth Colorimeter. The results of the QUV exposure of our UV cured films are shown in Table V. The results do not show a clear trend for either coating. While Coating C generally has lower ΔE values than Coating B, Coating C has a higher initial rate of color development. When the fluoropolymer resins are used in thermally cured systems improved resistance to yellowing is reported and good long term gloss retention is reported for UV cured films.[4] Typically, acrylates have ΔE values that range from 5.8 to 6.7 and the photoinitiator contributes to this yellowing.[16,17] What we have observed may be due to the photoinitiator. The addition of UV absorbers and hindered amine light stabilizers has been recommended to improve the performance of acrylates.[16,17]

Table V. Comparison of Yellowing vs Time Data for Urethane Acrylate Films

Time (Days)	ΔE Value (Based on white standard)	
	Coating B	Coating C
Initial Reading	6.27	5.29
1	7.18	7.82
2	7.66	8.08
5	8.42	8.20
8	7.89	6.96
13	8.78	7.51

A comparison of the initial DMA of the films of Coating B and Coating C with the DMA after seven days exposure to the QUV light is shown in Table VI. In spite of the higher functionality (2.4) of the FPR3UA the initial equilibrium modulus (E_0), which is proportional to the crosslink density, was lower for Coating B than for Coating C which contained the CN 963 with a functionality of only 2. For both coatings the E_0 decreases slightly, the T_g increases, and the temperature of the equilibrium modulus increases after QUV exposure. All by about the same magnitude, indicating no real advantage to the FPR3UA oligomer in Coating B. These QUV results could be affected by the presence of the TPGDA and other diluents should be evaluated.

Coating B and Coating C were also used to investigate the water sensitivity of the FPR3UA oligomer. Again 3 mil films cured at 2 J/cm^2 in nitrogen were used. A "Dynamic Water Soak Test" (Method A) was conducted over a 2 week period. A plot of the data is shown in Figure 5. Generally there is an initial absorption of water,

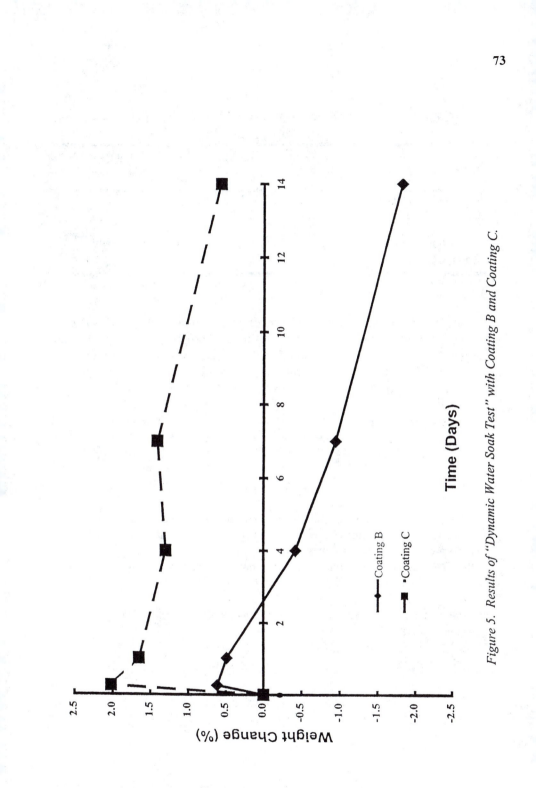

Figure 5. Results of "Dynamic Water Soak Test" with Coating B and Coating C.

Table VI. Dynamic Mechanical Analysis Data for Yellowed Free Radical Cured Films

Sample	T_g by Tan δ_{max} (°C)	E_o (MPa)	Temperature of E_o (°C)
Coating B No QUV	53.5	13.1	80.4
Coating B 10 Days QUV	70.0	12.5	98.9
Coating C No QUV	55.1	19.6	83.6
Coating C 10 Days QUV	71.0	18.1	102.9

indicated by the weight gain, followed by a decrease in film weight probably due to leaching of small molecules. From Method A the total weight lost for Coating B about 2.2% vs 1.3% for the Coating C control. Although no xylene odor was observed, the difference could be due to residual xylene solvent in the Coating B film. The maximum water absorption measured by Method A occurred after about 4 hours and was 1.4% for Coating B vs 2.3% for the Coating C control. Using Method B the Coating B absorbed 1.54% water vs a value of 1.62% for Coating C. These values can be compared with those of an optical fiber coating containing fluorine which gives a values of <0.6% up to 21 days by Method A.[13,14]

Conclusion

The refractive index of the base FPRs was determined to be about 1.46 and they were found to be compatible with several conventional acrylate diluents. Flexible SIPN's were formed from FPR/multifunctional acrylate blends but between 10 and 50 percent of the FPR was extracted from the cured film after 5 days in toluene. A fast curing urethane acrylate oligomer (functionality of 2.4) based on FPR3 was made. The cured FPRUA gave good physical-mechanical properties with yellowing equivalent to a conventional urethane acrylate and water resistance slightly better than the conventional urethane acrylate.

Acknowledgement

Thanks to DSM Desotech for the opportunity to publish this work, to M. Sullivan for technical assistance and to A. Tortorello and S. Lapin for their helpful discussions.

References

1. "Fomblin Perfluoro Polyethers", Technical Bulletin, Ausimont USA, Inc., 1992.

2. Simeone, G.; Turri, S.; Scicchitano, M; Tonelli, C. *Die Angewandte Makromolekulare Chemie,* 1996, 236, 111.
3. Marchionni, G.; Ajrold, G.; Pezzin, G. *Comprehensive Polymer Science,* Second Supplement, Aggarwal, S. L.; Russo, S., Eds.; Elsevier Science Inc.: Tarrytown, New York, 1997, Chapter 9, p. 347.
4. LFX-910LM, LF-916, LDG, CLG, (1994), and "Lumiflon Fluoropolymer Resins", Technical Bulletins, Zeneca Resins, 1986.
5. Priola, A.; Bongiovanni, R.; Malucelli, G.; Pollicino, A.; Tonelli, C. *Macromol. Chem. Phys.* 1997, 198, 1893.
6. R. Bongiovanni, R., G. Malucelli, G., A. Pollicino, A., C. Tonelli, C., G. Simeone, G. and A. Priola, A. *Macromol. Chem. Phys.* 1998, 199, 1099.
7. Head, R. A; Fitchett, M. *RadTech '88 - NA Conference Procedings,* April 24-28, 1988, New Orleans, LA, p 309.
8. Okamoto, S.; Miyazaki, N. (Asahi Glass), Kokai Tokkyo Koho JP 01,051,418 (8/21/87).
9. Miura, R.; Kodama, S. (Asahi Glass), Japan Kokai Tokkyo Koho JP 05,279,435 (10/26/93).
10. Munakata, S.; Unoki, M. (Asahi Glass), Kokai Tokkyo Koho JP 61,296,073 (6/25/85).
11. Miura, R.; Kodama, S. (Asahi Glass), Japan Kokai Tokkyo Koho JP 08,319,455 (12/3/96). CA 126: 158848n.
12. Noren, G. K. *Journal of Coatings Technology,* in press.
13. Szum, D. *Proceedings of the 3rd Nürnberg Congress,* paper 26, March 13-15, 1995.
14. Szum, D. *European Coatings Journal,* 1995, No. 9, 254.
15. *Polymer Handbook,* 2nd Ed., Brandrup, J.; Immergut, E. H., Eds.; Wiley and Sons, New York, 1975, p III-241.
16. Cauffman, T. E. *Modern Paint and Coatings,* June 1995, p 32.
17. Yang, B. *Modern Paint and Coatings,* May 1996, p 40.

Chapter 6

Synthesis and Characterization of Novel Chiral Chromophores for Nonlinear Optical Applications

Shuangxi Wang[1], Kenneth D. Singer[2], Rolfe G. Petschek[2], S. Huang[3], and L.-C. Chien[1],*

[1]Liquid Crystal Institute and [3]Department of Chemistry, Kent State University, Kent, OH 44242
[2]Department of Physics, Case Western Reserve University, Cleveland, OH 44106

We report the design, synthesis and characterization of novel conjugated chiral materials, based on the derivatives of camphorquinone as shown below, for nonlinear optics. Several conjugated chiral monomers have been prepared and characterized. A x-ray single crystal study was also performed on one of the chiral monomer. The chiral conjugated polymers were prepared from polycondensation of conjugated chiral diol-monomer with di-functional and multifunctional co-monomers. We present results of synthesis and characterization of the conjugated chiral materials.

Introduction

Recently, advances in optical technology have attracted great interest in the fabrication of second-order nonlinear optical (NLO) devices for frequency conversion and electro-optic modulation. Photorefractive organic polymers are emerging as key materials for advanced information and telecommunication technology.[1,2] Second-order NLO polymers, polymers which are functionalized with second-order NLO chromophores, offer advantages of large susceptibilities, high laser damage threshold, faster response time, a high density of NLO moieties, versatility of molecular structural modifications, improved processability and relatively higher thermal stability compared with small molecular weight organic materials.[3]

77

In electro-optical polymers, a commonly used figure of merit of chromophores for optoelectronic applications is the product of first hyperpolarizability of the molecule, β, and the dipole moment, μ. The way to maximize the $\beta\mu$ product in chromophores is by adjusting the strength of the donor and acceptor substitutents. Recently, we reported the octopolar contribution to β^x and other off-diagonal contributions in the Cartesian tensor,[4] contrary to the traditions picture of a dominantly one-dimensional doplar charge-transfer axis. If the π electrons in a molecule are conjugated in one dimension, Kleinman symmetry is automatic. Thus, molecules must feature electrons that are delocalized in two or three dimensions to be considered good candidate materials in which a violation of Kleinman symmetry can be examined. Molecules that exhibit chiral symmetries and excitations are examples of multidimensional molecules that would be expected to exhibit three dimensional delocalization and Kleinman symmetry breaking.

In this paper, we report the design, synthesis and characterization of novel elastic-polymers containing NLO chiral conjugated chromophores, based on the derivatives of chiral camphorquinone, which could be expected to display large three dimensional polarity and act as more thermal stable NLO optical materials.

Experimental

The starting materials were purchased from Aldrich Chemical Company Inc. and were used without purification. Column chromatography was performed on silica gel (Aldrich Chemical Company, Inc., 70-230 mesh, 60 Å). ^{1}H NMR (200 MHz) spectra were recorded on a Varian Gemini-200 spectrometer. IR spectra were recorded on a Nicolet Magna_IR 550 Spectrometer. Differential scanning calorimetry (DSC) of the polymers was performed using a Perkin Elmer DSC system equipped with 7/DX thermal analysis controller. The experiments were conducted under nitrogen atmosphere; the scanning rates was 5 °C/min in all cases. Gel permeation chromatography (GPC) was performed on a waters 510 HPLC instrument with a Waters 410 differential refractometer. The experiments were done using THF as solvent (1 mL/min, 35 °C), with polystyrene standards. For X-ray single crystal structure analysis of a NLO chromophor, the crystal was grown from ethyl acetate/n-hexane by slow evaporation at room temperature. Cell parameters and intensity data were derived from measurement on four-circle diffractometer-Rigaku ACF5R. Molecular and crystal structures were determined by the direct method by using the program of TEXSAN.

Synthesis of 4-nitro-(4-N, N'-diallylphenylazo)benzene (1). To a solution of 4-(4-nitro phenylazo)aniline (4.84g, 0.02mol) and sodium hydroxide (1.6g, 0.04 mol) in 80 mL of DMF was added allyl bromide (10g , 0.08mol) stirring at room temperature. The dark-red solution obtained was allowed to stir at room temperature for 24 hours. The reaction mixture was poured into 500 mL of water to give an organic solid which was collected by filtration and purified by chromatography using hexane/ethyl acetate (1:3/v:v) as an eluent to yield 5.8 g of 4-nitro-(4-N, N'-diallylphenylazo)benzene (1) (yield: 93%). ^{1}HNMR (CDCl$_3$): δ 8.32 (2H, d, J = 8.2, Ar-H), 7.92 (2H, d, J = 8.0, Ar-H), 7.88 (2H, d, J = 8.4, Ar-H), 7.772 (2H, d, J = 8.4,

Ar-H), 5.80-5.98 (2H, m, Vinyl-H), 5.16-5.26 (4H, m, Vinyl-H), 4.06-4.04(4H, m, 2(-CH$_2$-). IR (cm^{-1}, KBr pellet): 1335, 1258 (NO$_2$).

Synthesis of 4-(4-N, N'-diallylphenylazo)aniline (2). A mixture of 4-nitro-(4-N, N'-diallylphenylazo)benzene (**1**) (5g, 0.0155 mol) and sodium sulfite nonahydrate (10g, 0.042 mol) in 100 mL of Ethanol was refluxed for 12 hours. The resultant deep-red solution was poured into stirring ice-water (~500 g) to give red solid which was applied to chromatography using hexane/ethyl acetate (2:1/v:v) as an eluent. 3.6g of 4-(4-N, N'-diallylphenylazo)aniline (**2**) was obtained as a dark-red crystal (yield: 79%). ^{1}HNMR (CDCl$_3$): δ 7.86 (2H, d, *J* = 9.0, Ar-H), 7.78 (2H, d, *J* = 9.2, Ar-H), 6.76 (2H, d, *J* = 8.2, Ar-H), 6.72 (2H, d, *J* = 8.0, Ar-H), 5.81-5.98 (2H, m, Vinyl-H), 5.16-5.26 (4H, m, Vinyl-H), 4.05-5.26 (4H, m, 2(-CH$_2$-). IR (cm^{-1}, KBr pellet): 3334 (NH$_2$).

Synthesis of monomer 1. A solution of 4-(4-N, N'-diallylphenylazo)aniline (**2**) (572 mg, 1.96 mmol), S-(+)-camphorquinone (326 mg, 1.96mmol) and 2-3 drop of acetic acid in 30 mL of toluene was refluxed and water was removed through a Dean-Stark trap to dry the system. After reflux for 24 hours, the solvent was removed by rotary evaporation. Resultant residue was recrystallized from a mixture of ethyl acetate and hexane to give pure monomer **1** as a dark-brown crystalline substance. Large, well shape crystals suitable for X-ray diffraction were obtained by slow evaporation of a solution of monomer **1** in ethyl acetate/n-hexane. ^{1}HNMR (CDCl$_3$): δ 7.85 (2H, d, *J* = 8.4, Ar-H), 7.85 (2H, d, *J* = 8.8, Ar-H), 6.84 (2H, d, *J* = 8.8, Ar-H), 6.77 (2H, d, *J* = 9.0, Ar-H), 5.79-5.95 (2H, m, Vinyl-H), 5.15-5.22 (4H, m, Vinyl-H), 4.01-4.02 (4H, m, 2(-CH$_2$-), 1.25-2.88 (5H, m, camphorquinone-H), 0.99 (3H, s, camphorquinone-Me), 0.98(3H, s, camphorquinone-Me), 0.93 (3H, s, camphor quinone-Me). IR (cm^{-1}, KBr pellet): 1750 (C=O), 1681 (C=N).

Synthesis of polymer P1NO$_2$. A mixture of 4-nitro-(4-N, N'-diallylphenylazo)benzene (**1**) (2g, 6.204 mmol), 1, 1', 3, 3'-tetramethyldisiloxane ((0.833g, 6.204 mmol) and 3 drops of the catalyst platinum(0)-1,3-dDivinyl-1,1,3,3-tetramethyldisiloxane in xylenes was melted with stirring at 90 °C overnight under a N$_2$ atmosphere and then poured into methanol. The precipitate was collected, redissolved in dichloromethane, and filtered to remove the catalyst residue. The filtrate was concentrated and precipitated into methanol, followed again by filtration and reprecipitation. ^{1}HNMR (CDCl$_3$): δ 8.30 (2H, d, *J* = 8.2, Ar-H), 7.90 (4H, m, Ar-H), 6.70 (2H, Ar-H), 3.35-4.04 (4H, m, 2(-CH$_2$-)), 0.41-1.66 (4H, m, 2(-CH$_2$-)). IR (cm^{-1}, KBr pellet): 1338, 1258 (-NO$_2$).

Synthesis of Polymer P1NH$_2$. A mixture of **P1NO$_2$** (1 g) and sodium sulfite nonahydrate (5g) in 60 mL of THF was refluxed and the solution became dark-red. After heated to reflux for 5 hr, the solvent was removed by rotary evaporation to give a dark-red oil residue, which was treated with water-dichloromethane (50-10 mL). The organic phase was collected and evaporated to dryness. The resultant dark-red oil was dried under vacuum and used in the next step without purification. ^{1}HNMR (CDCl$_3$): δ 7.88 (2H, d, *J* = 9.0, Ar-H), 7.79 (2H, d, *J* = 9.2, Ar-H), 6.76 (2H, d, *J* = 8.2, Ar-H), 6.72 (2H, d, *J* = 8.0, Ar-H), 3.35-4.04 (4H, m, 2(-CH$_2$-)), 0.41-1.66 (4H, m, 2(-CH$_2$-)). IR (cm^{-1}, KBr pellet): 3354 (NH$_2$).

Synthesis of Polymer 1. To a solution of **P1NH$_2$** (195 mg), S-(+)-camphorquinone (500 mg, 3.00 mmol) in dry 50 mL of toluene were added to 0.5 mL of TiCl$_4$ by syringe

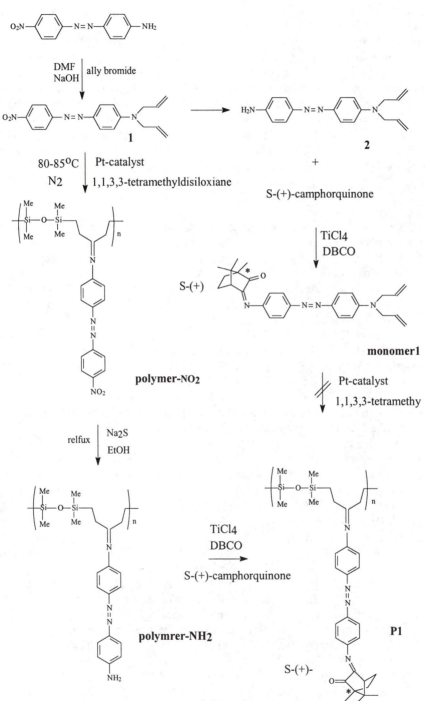

Scheme 1

and 1,4-diazabicyclo[2.2.2]octane (0.5 g). The mixture was refluxed for 16 hours and allowed to cool to room temperature and the insoluble solid was filtered off and washed with toluene. The filtrate was concentrated and precipitated into methanol, followed again by filtration and reprecipitation. 110 mg of polymer 1 was obtained as deep-red precipitate. ^{1}HNMR (CDCl$_3$): δ 7.87 (2H, Ar-H), 7.82(2H, Ar-H), 6.81(2H, Ar-H), 6.76(2H, d, J = 9.0, Ar-H), 3.38-4.06 (4H, m, 2(-CH$_2$-)), 0.41-1.68(4H, m, 2(-CH$_2$-), 1.25-2.88(5H, m, camphorquinone-H), 0.99 (3H, s, camphorquinone-Me), 0.98(3H, s, camphorquinone-Me), 0.93(3H, s, camphorquinone-Me). IR (cm^{-1}, KBr pellet): 1750 (C=O), 1682 (C=N).

Synthesis of Compound 2a. To 250 mL of three necked, round-bottomed flask fitted with a dry nitrogen inlet, septum, reflux condenser and mercury seal, 4-nitro-benzyltriphenylbenposphonium bromide (4.673g, 0.01mol), was introduced with additional funnel and the solid was suspended in try toluene (50 mL). To the suspension was added potassium tert-butyloxide (1.1g, 0.01 mol). The dark red solution was heated at 80 °C for 1hr and then S-(+)-camphorquinone (1.62g, 0.01 mol) was added. The resultant mixture was refluxed for 8hr and then diluted with benzene (50 mL) and brine (100 mL). The organic layer was separated, washed and dried. The crude solid was obtained after removal of solvent and was further purified by column chromatography using hexane/ethyl acetate (7:1/v:v) as the eluent. After recrystallization from ethanol produced a near white crystalline compound 2a (1.6 g, yield: 56%). ^{1}HNMR (CDCl$_3$): δ 8.25 (2H, d, J = 8.8, Ar-H), 7.60 (2H, d, J = 8.8, Ar-H). 7.27 (1H, s, C=C-H), 3.08 (1H, d, J = 4.4, -CH in camphorquinone unit), 1.53-2.24 (4H, m, -CH$_2$ in champhorquinone unit), 1.06(3H, s, Me-camphorquinone), 1.03 (3H, s, Me-camphorquinone), 0.81 (3H, s, Me-camphorquinone). IR (cm^{-1}, KBr pellet): 1750 (C=O), 1338 (NO$_2$).

Synthesis of Compound 2b. A mixture of 2a (1.8g, 0.063 mol) and tin(II) chloride (5. 4g) in 40 mL of concentrated hydrogen chloride and 40 mL of ethanol was refluxed for three hours. The resulting dark-red solution was cooled to room temperature and ethanol was removed by rotary evaporation. The resulting aqueous solution was chilled to offer 2b·xHCl as a dark-red solid (1.63g), which was used in next step without further purification. %). IR (cm^{-1}, KBr pellet): 3358 (NH$_2$), 1752 (C=O).

Synthesis of Monomer 2d. 2b·xHCl (1.48 g) was suspended in a mixture of glacial acetic acid (10mL) and concentrated hydrochloride acid (10 mL). The resulting mixture was heated for 10 minutes and then sodium tetrafluoroborate (1.2g) in 5mL of water was added and cooled to 0~5 °C in ice-bath. Sodium nitrite dissolved in water was cooled to 0 °C and then dropwise added to the suspension of 2b·xHCl until end point was reached as indicated by iodine-starch paper that became dark-violet. The resulting solution was stirring at 0 °C for 5 minutes and N-phenyldiethanolamine (1.1g, 0.06 mol) dissolved in acetic acid was added in one portion. The resulting mixture was stirring at room temperature for 2 hours and filtrated to remove insoluble solid. The dark-red filtrate was neutralized by aqueous sodium hydroxide solution and the red solid obtained was purified by chromatography using ethyl acetate as the eluent and the obtained crude was recrystalized from hexane/ethyl acetate to yield 780 mg of the monomer 2 as a orange-red solid. %). ^{1}HNMR (CDCl$_3$): δ 7.87 (2H, d, J = 8.0, Ar-H), 7.86 (2H, d, J = 8.6, Ar-H), 7.58 (2H, d, J = 8.6, Ar-H). 7.27 (1H, s, C=C-H), 6.74 (2H, d, J = 8.8, Ar-

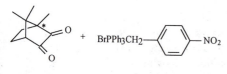
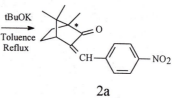

2a

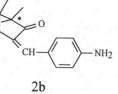
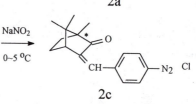

2b

2c

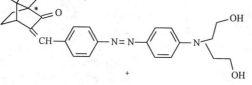

Monomer 2

+

OCN—(CH₂)₆—NCO

DMF
60~80 °C | triethylamine

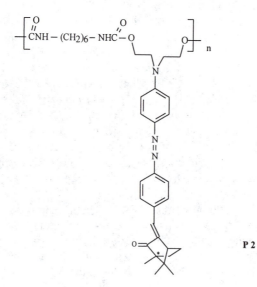

P 2

Scheme 2

H), 3.87-3.93 (4H, m, 2(-CH$_2$-)), 3.70-3.75 (4H, m, 2(-CH$_2$-)), 3.27 (2H, Broad, OH-), 3.17 (1H, d, J = 4.4, -CH in camphorquinone unit), 1.53-2.24 (4H, m, -CH$_2$ in champhorquinone unit), 1.05 (3H, s, Me-camphorquinone), 1.02(3H, s, Me-camphorquinone), 0.82(3H, s, Me-camphorquinone). IR (cm^{-1}, KBr pellet): 1725 (C=O).

Synthesis of 3a. A mixture of **2a** (2g, 0.007 mol) and Lawesson reagent (3.5g, 0.008 mol) in 150 mL of anhydrous toluene was refluxed under a N$_2$ atmosphere for 24 hours. The dark green solution obtained was cooled to room temperature and the solvent was removed by rotary-evaporation. The resultant residue was soluble in a minimum amount of ethyl acetate and applied to a sillical gel using hexane/ethyl acetate 8:1/v:v as the eluent. The product **3a** as a dark green crystalline was obtained upon removal of the solvents (1.2g, 76%). ^{1}HNMR (CDCl$_3$): δ 8.26 (2H, d, J = 9.0, Ar-H), 7.66 (2H, d, J = 8.8, Ar-H). 7.55 (1H, s, C=C-H), 3.21 (1H, d, J = 4.4, -CH in camphorquinone unit), 1.53-2.30 (4H, m, -CH$_2$ in champhorquinone unit), 1.23 (3H, s, Me-camphorquinone), 1.10 (3H, s, Me-camphorquinone), 0.77 (3H, s, Me-camphorquinone). IR (cm^{-1}, KBr pellet): 1712 (C=S), 1338 (NO$_2$).

Synthesis of 3b. A mixture of **3a** (1.02g, 0.0034 mol), HgO (0.7361g, 0.0034 mol), and malononitrile (220 mg, 0.0034 mol) was dissolved in anhydrous ethanol and three drops of piperpyridine was added to the mixture as a catalyst. The mixture was refluxed for 12 hours. The dark-red mixture was cooled to room temperature and filtrated to remove insoluble solid. The filtration was evaporated to dryness under rotary evaporation to offer dark-red solid which was purified by chromatography (sillical gel) using hexane/ethyl acetate (2:1/v :v). The product **3b** was obtained as yellow-orange solid (450 mg, yield 46%). ^{1}HNMR (CDCl$_3$): δ 8.41 (2H, d, J = 8.8, Ar-H), 7.55 (2H, d, J = 8.8, Ar-H). 6.03 (1H, s, C=C-H), 3.21 (1H, d, J = 4.4, -CH in camphorquinone unit), 1.55-2.32 (4H, m, -CH$_2$ in champhorquinone unit), 1.26 (3H, s, Me-camphorquinone), 0.90 (3H, s, Me-camphorquinone), 0.85(3H, s, Me-camphorquinone). IR (cm^{-1}, KBr pellet): 2222 (C≡N), 1336(NO$_2$).

Synthesis of 3c. A mixture of **3b** (2g) and sodium sulfite nonahydrate (5g, 0.021 mol) in 100 mL of ethanol was gently refluxed for 1 hour. The solvent was removed and the resultant residue was treated by water / dichloromethane (100 mL/100mL). The organic phase was dried and evaporated by rotary evaporation. A dark-red solid was obtained and used in next step without further purification. ^{1}HNMR (CDCl$_3$): δ 7.48 (2H, d, J = 8.8, Ar-H), 6.98 (2H, d, J = 8.8, Ar-H). 6.03 (1H, s, C=C-H), 3.21 (1H, d, J = 4.4, -CH in camphorquinone unit), 1.55-2.32 (4H, m, -CH$_2$ in champhorquinone unit), 1.26 (3H, s, Me-camphorquinone), 0.90 (3H, s, Me-camphorquinone), 0.85 (3H, s, Me-camphorquinone).

Synthesis of Monomer 3. The synthetic procedure for monomer **3** was the same as that for monomer **2** (yield: 58%). ^{1}HNMR (CDCl$_3$): δ7.86 (2H, d, J = 8.0, Ar-H), 7.85 (2H, d, J = 8.6, Ar-H), 7.78 (2H, d, J = 8.6, Ar-H). 6.89 (2H, d, J = 8.9, Ar-H), 6.03 (1H, s, C=C-H), 3.87-3.93 (4H, m, 2(-CH$_2$-)), 3.70-3.754H, m, 2(-CH$_2$-)), 3.27 (2H, Broad, OH-), 3.17 (1H, d, J = 4.4, -CH in camphorquinone unit), 1.53-2.24 (4H, m, -CH$_2$ in champhorquinone unit), 1.05 (3H, s, Me-camphorquinone), 1.02 (3H, s, Me-camphorquinone), 0.82 (3H, s, Me-camphorquinone). IR (cm^{-1}, KBr pellet): 2217 (C≡N). *Synthesis of polymer P2*. A mixture of monomer **2** (178mg, 0.4 mmol) and 1,6-diisocyanohexane (68 mg, 0.4 mmol) in 5 mL of DMF was heated for 10

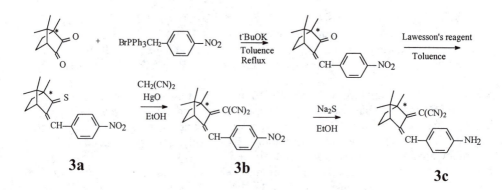

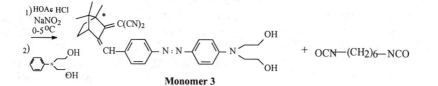

Monomer 3

$+ \quad OCN-(CH_2)_6-NCO$

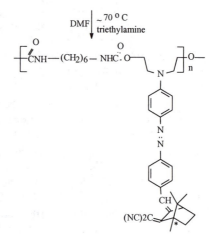

P3

Scheme 3

minutes at 30°C under N_2 atmosphere, then three drops of triethylamine was added as a catalyst. The resulting mixture was heated at 80~100 °C for 24 hours, and poured into methanol to produce red precipitate. The precipitate was collected, redissolved in dichloromethane, and filtrated. The filtrate was concentrated and precipitated into methanol, followed again by filtration and reprecipitation (210 mg, yield: 72%). ^{1}HNMR (CDCl$_3$): δ 7.86 (4H, m Ar-H), 7.54 (2H, d, J = 8.6, Ar-H), 6.89 (2H, Ar-H). 7.27 (1H, s, C=C-H), 5.56 (2H, broad, N-H), 3.32-4.44 (9H, m, 4(-CH$_2$-) and -CH in camphorquinone unit), 1.33-2.23 (16H, m, (-C$_6$H$_{12}$-) and -CH$_2$ in champhorquinone unit), 1.05 (3H, s, Me-camphorquinone), 1.02 (3H, s, Me-camphorquinone), 0.70(3H, s, Me-camphorquinone). IR (cm^{-1}, KBr pellet): 1716 (C=O), 1730(C=O).

Synthesis of polymer P3. The synthetic procedure for polymer **3** was the same as that for the polymer **2** (dark-red solid, yield: 68%). ^{1}HNMR (CDCl$_3$): δ 7.88 (4H, m Ar-H), 7.53 (2H, d, J = 8.6, Ar-H), 6.89(2H, Ar-H). 6.10 (1H, s, C=C-H), 5.80 (2H, broad, N-H), 3.32-4.44 (9H, m, 4(-CH$_2$-) and -CH in camphorquinone unit), 1.33-2.23 (16H, m, (-C$_6$H$_{12}$-) and -CH$_2$ in champhorquinone unit), 1.05 (3H, s, Me-camphorquinone), 1.02 (3H, s, Me-camphorquinone), 0.70 (3H, s, Me-camphorquinone). IR (cm^{-1}, KBr pellet): 2218 (C≡N), 1730(C=O).

Results and Discussion

Synthesis of monomers and polymers. Monomer **1**, polymers **1** were synthesized according the Scheme1, and characterized by ^{1}NMR, IR UV-vis, DSC and GPC. The physical properties of these materials are summarized in Table 1. We first attempted to use monomer **1** to polymerize with tetramethyl-disiloxane in the presence of platinum (Pt) catalyst at 80-110 °C. However, under the high temperature reaction conditions, we were unsuccessful to obtain the desired polymer because partial decomposition of the monomer obscured isolation of the product. Apparently, oxygen in the camphorquinone and the nitrogen of the Shiff's base of monomer **1** coordinated to Pt form a chelate complex. In addition, cyclization occurred for acyclic diene monomer **1** and tetramethydiloxane forming small molecular cyclic compound at a low concentration of reactants. To avoid these problems,direct condensation polymerization was performed on intermediate **3** with tetramethyl disiloxane without solvent at temperature 80-85 °C (melting point for **3**). The polymerization reaction was catalyzed by Pt-DVTMDS to afford polymer **1** in good yield (80%). The nitro functional group of polymer **1** was reduced to amine to give polymer **2**. Condensation of polymer **2** with chiral camphorquinone produces polymer **3**. According to DSC results, both polymers **1** and **2** exhibit low glass transition temperatures. This is not surprising due to the flexible nature of the polymer backbones. In addition, low molecular weight polymers are commonly obtained. An x-ray single crystal structure analysis of monomer **1** is shown in Figure 1. The monomer **1** crystallizes in a non-centrosymmetric monoclinic space group P_{21} (#4) with a=10.7814(9) Å, b=10.2256(9) Å, c=22.0652(19) Å, β=97.955(2)°, V = 2409.2(4) Å^3, Z = 4 , R= 6.54 % and R$_w$

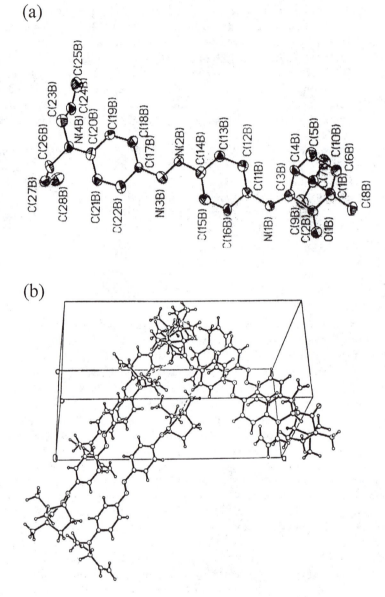

Figure 1. (a) The x-ray single crystal structure, and (b) the molecular structure of monomer **1d**. The compound **1d** crystallizes in monoclinc asymmetric center spacer group P_{21}.

=12.76, although the molecules in unit cell are aligned in two dimensions due to the strong dipole-dipole interactions as shown in Figure 1.

In order to produce more stable NLO polymer, we synthesized polymer **2**, in which the connection to the chiral unit was changed to a C=C double bond from a C=N double bond in polymer **1**. The synthesis of the polymer **2** is summarized in Scheme 2. The reaction of chiral camphorquinone with (4-Nitrobenzyl)triphenyl phosphnium bromide by Witting reaction using potassium tert-butoxide as base at refluxing of toluene gave **2a** in yield 48%. Reduction of **2a** by SnCl$_2$ and HCl gave **2c** (yield: 92%). Diazotiation of **2c** using nitrous acid, followed by azo-coupling with N-phenyldiethanolamine afforded **monomer 1** in 52 %yield. Polymerization of **monomer 1** with 1,6-diisocyanatohexane using triethylamine as catalyst in dry DMF at 60-80 ºC afforded the desired **polymer 2** (yield 56%). We also synthesized **P3** based on **P2,** in which one oxygen in chiral unit was substituted by dinitrile groups, in order to enhance electronic conjugation between electronic donating group and electronic accepting group. The completion of the synthesis of **P3** is outlined in Scheme 3. First of all, our attempts to condense camphorquinone directly with malononitrile by Knoevenagel reaction to get **3b** was unsuccessful. So, we prepared first a thiocarbonyl intermediate **3a** by the reaction of camphorquinone with Lawsson's reagent and then Knoevenagel condensation with malononitrile mediated by HgO gave the target **3b**. Reduction of **3b** by Na$_2$S (in ethanol, gently refluxing for 1 hour), followed by diazotiation, azo-coupling gave **monomer 3**. Polymerization of monomer **3** with 1,6-diisocyanatohexane using the same reaction condition for **P2** afforded **P3**.

Characterization. Monomers **1c** with extended π conjugation, compared with *p*-nitroanaline, is a sunstantial enhancement of the Kleinman allowed components and a moderate increase in Kleiman disallowed components.[5] The synthesized polymers **P1-3** were characterized by UV, IR, NMR, DSC, TGA and GPC. In the IR spectra of camphorquinone, monomer **1** and polymer **P1**, the doublet peaks at 1769 and 1760 cm-1 for C=O groups in the IR spectrum of the camphorquinone become single peak in the IR spectra of monomer **1** and polymer **1** due to the condensation of one carbonyl group in monomer **1** and Polymer **1**. For the infrared spectra of polymers **3** and **4**, polymer **3** shows a peak at 2228 cm-1 due to -CN groups which is absent for polymer **2**. In the ^{1}H NMR spectra of monomer**1**, polymer-**NO$_2$** and polymer **1**, the monomer **1** shows the vinyl around 5.6 to 6.0 pm, which disappeared in both the ^{1}H NMR spectra of polymer-**NO$_2$** and polymer **1**. The peaks of methyl group of siloxane appear in polymer **1** and polymer-**NO$_2$** in range from 0.0 to 0.4 ppm, which are absent in monomer **1**. Polymer **1** and monomer **1** show three characteristic peaks of methyl groups of camphorquinone in the range from which are absent in polymer-**NO$_2$**. The peaks of the^{1}H NMR of polymer **1** and polymer-**NO$_2$** are broad and low resolution compared with that of small molecule of monomer **1**, typical of polymer spectra. The Uv-vis spectra of polymers **P1**, **P2** and **P3**, as shown in Figure 2 were measured in CH$_2$Cl$_2$. The maximum absorbance for polymers **P1**, **P2** and **P3** is 413, 438 and 490 nm, respectively. If we compare values of the absorbance, one conclusion can be drawn from this data: the bathochromic shifted is a manifestation of the push-pull effect, namely, the maximum absorbencies are red-shifted as electronic conjugation increases from polymer **P1** to **P3**. The

Table 1. The physical properties of polymers P1, P2 and P3.

Compound	$[\alpha]^{20}$	Tg/m.p. (°C)	Td_5 (°C)	Mw	MWD
P1	38	56	257	10,200	1.56
P2	40	56	279	7,200	1.27
P3	39	69	283	8,700	1.41

$[\alpha]^{20}$ = optical rotational power; Tg = glass transition temperature; Td_5 = decomposition temperature of 5% weight loss in air by TGA; m.p. = melting point; MWD = M_w/M_n.

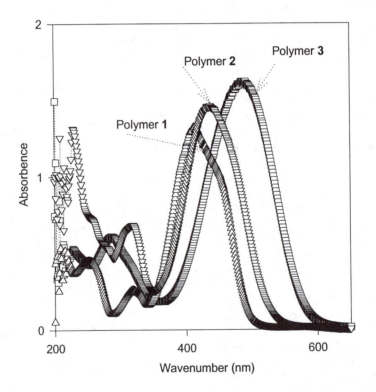

Figure 2. The absorption spectra for polymers P1, P2 and P3.

physical properties of polymer **P1-3** are summarized in Table 1. As seen in this table, the present polymers **1-3** exhibit relatively close in optical rotation power values. These low-molecular-weight polymers (Mw = 7200 ~10200) have moderate glass transition temperatures, which are not surprising because of the soft nature of the polymer backbones. Polymer **P3** has a higher glass transition temperature than those of the **P1** and **P2** perhaps due to the increase in rigidity from the extended molecular conjugation. These polymers exhibit high thermal stability with a 5% weight loss temperature ranging from 257~283 °C measured by the thermal gravimetric analyses. Polymers **1-3** have large optical rotation power, thus these materials are expected to possess a larger SHG.

Conclusions

Three polymers containing NLO chromophores of chiral champhorquinone derivatives were synthesized and characterized. The crystal structure of monomer **1**, determined by X-ray diffraction, was found to crystallize in monoclinic unsymmetric center space group. Monomer 1c was found to exhibit large Kleinman allowed component and moderate Kleiman disallowed component studied by the hyper-Rayleigh scattering method.[6] The UV absorption of chiral conjugated polymers are red-shifted with the increasing in electronic conjugation. The study of NLO properties of the polymers is in progress.

This research was supported by NSF ALCOM grant #DMR 89-20147.

References

1. G. A. Lindsay and K.D. Singer, "Polymers for Second-Order Nonlinear Optics," ACS Symposium Series #601, Wingshington DC, 1995.
2. H.S. Nalwa and S. Miyata," Nonlinear Optics or Organic Molecules and Polymers, CRS Press, Boca Raton, FL 1996.
3. S. A. Jenkhe and K. J. Wynne, " Photonic and Optoelectronic Polymers" ACS Symposium Series #672, Wingshington DC, 1997.
4. S. F. Hubbard, R. G. Petschek, K. D. Singer, N. D'Sidocky, C. Hudson, L.C. Chien, C.C. Henderson and P. A. Cahill, J. Opt. Soc. Am. B **15**, 298-30, 1998.
5. (a) S.J. Lalama, A.F. Garito, *Phys. Rev. A* **20**, 1179 (1979). (b) J. Zyss, T.C. Van, C. Dhenaut and I. Ledoux, *J. Chem. Phys.***177**, 281 (1993). (c) S. Sadler, R. Dietch, G. Bourhill, and Ch. Brauchle, *Opt. Lett.* **21**, 251 (1996).
6. V. Ostroverkhov, O. Ostroverkhov, R. G. Petschek, K.D. Singer, L. Sukhomlinova, R. J. Twieg, S.X. Wang, L.C. Chien, *Chemical Physics* **257**, 263 (2000).

Chapter 7

Miscibility Investigation of Fluorocarbon Copolymer and Methacrylate Copolymer Blends

Melynda C. Calves and J. P. Harmon

Department of Chemistry, University of South Florida, 4202 East Fowler Avenue, Tampa, FL 33620–5250

Fluorocarbon copolymer blends are studied for use as low refractive index cladding materials. The limits of phase separation for these blends were determined using both thermal and optical measurements. This study on optically clear vinylidene fluoride/tetrafluoroethylene copolymer and methyl methacrylate/ethyl acrylate coploymer blends reports UV/VIS spectroscopy, differential scanning calorimetry (DSC), dynamic mechanical analysis (DMA), dielectric analysis (DEA), fourier transform infrared spectroscopy (FTIR), and refractive index data.

Fluorocarbon methacrylate blends are important transparent materials for applications in plastic optical fibers (POFs). Early studies focused on the miscibility of blends composed of methacrylate and vinylidene fluoride homopolymers[1,2,3,4,5].

Later, blends of methacrylate copolymers and fluorocarbon copolymers were shown to exhibit a greater miscibility range than that of the homopolymer blends[6,7]. These copolymer blends are excellent choices as cladding materials because they possess low refractive indices and high glass transition temperatures [7]. Neat fluorocarbon polymers possess many advantages including chemical resistance, resistance to atmospheric corrosives, impact resistance and abrasion resistance [8]. These neat fluorocarbons are semi-crystalline in nature and are opaque in the UV visible region of the electromagnetic spectrum.

Past experiments include UV/VIS[7], FTIR[9], and refractive index data[6,7] showing the miscibility range for the homopolymer PVDF/PMMA blends[4-7]. FTIR spectra of the homopolymer blends has designated specific interactions between the carbonyl group of the PMMA and the ethylene group of the PVDF [9].

Thermal measurements have also classically been used to determine miscibility in blends [1,7,10]. The presence of an intermediate T_g between those of the two pure components constituting a blend indicates miscibility, while phase separation is identified by the emergence of two T_gs [11]. Masking of these two separate relaxations may, however, occur when crystalline components suppress the glass transition.

DSC data for the homopolymer blends have shown this masking effect of two separate T_gs at intermediate concentrations[12]. Although, increasing concentration of the crystalline component to well above 50% by weight reveals a second T_g [7,13,5].

Other indications of compatibility are exhibited by DMA [4,13]. Previous dynamic mechanical studies have shown two transitions for PMMA, an α and a β. The β mechanism involves motion of the side chain ester groups, and the α mechanism involves large scale motion associated with the glass transition temperature [4]. Four different transitions have been observed for pure PVDF. A lower temperature γ corresponding to restricted motion in the amorphous regions, a β transition corresponding to the glass transtion, a γ' associated with folding movements in the amorphous portions, [4] and an α transition attributed to local motion in the crystalline regions[14] have been reported. Mechanical studies of blends of PVDF/PMMA have exhibited only one T_g for high concentrations of crystalline material, not the expected two T_g's for phase separated blends [4,14]. The amorphous fraction of the blend is so low in concentration that a T_g is not detected.

Viscoelastic relaxations indicate miscibility via molecular motion and interaction [10,12.8]. Several relaxations are detected in DEA spectra for PVDF/PMMA blends. The lowest temperature transition, α_a, is attributed to segmental motion in the amorphous interface. The second transition,α_c, is due to the amorphous region of the crystalline PVDF. The high temperature transition, α_m, results from the molecular motions in the miscible phase of PVDF/PMMA. Also, a β relaxation is attributed to local motion in the amorphous phase [12,8]. Relaxations in pure PMMA include a low temperature β corresponding to side group rotation and an $\alpha\beta$ transition corresponding to T_g [10].

Previous studies of the homopolymer PMMA/PVDF blends have focused on characterization with a limited study of the onset of phase separation. The following study focuses on the limits of phase separation for PMMA/ethyl acrylate, PVDF/PTFE copolymer blends. Optical measurements, as well as thermal measurements, were utilized in this miscibility investigation. Any use of these blends in fiber cladding applications precludes the onset of crystallization and opacity. It is, therefore, important to probe the blend matrix by the various methods used here in.

Experimental

Materials and Sample Preparation

Kynar SL, a random copolymer of 80% vinylidene difluoride and 20% tetrafluoroethylene was obtained from Elf-Atochem. CP-41, a copolymer of 90%

methyl methacrylate and 10% ethyl acrylate, was obtained from Continental Polymers. Samples of Kynar SL/CP-41 blends were prepared with varying concentrations by weight percent Kynar from 0% to 100%. The blends were prepared by dissolving the copolymers in acetone at a concentration of 80% by weight. After evaporation in air, each of the samples was placed in a vacuum oven at 55°C for 48 hours to ensure complete removal of acetone. FTIR data confirms the removal of the acetone.

Analytical Techniques

UV/VIS Spectroscopy

Thin film disks were formed by pressing cylindrical dyes of diameter 24.4mm wide and 0.8mm thick between two ferrotyping plates in a hot press at 200°C with a pressure of 5 metric tons. These disks were placed between two quartz glass plates coated with PDMS oil to provide optical contact. Transmission spectra were recorded on a Hewlitt-Packard Model 8452A Spectrophotometer.

Refractive Index and Numerical Aperture

Refractive indices were obtained from 1.6mm thick samples using an Abbe' refractometer. Samples were coated with polymethylphenyl siloxane oil with a refractive index of 1.5489.

Fourier Transform Infrared Spectroscopy

Thin film samples were prepared by first dissolving blends in acetone, then using 0.4mm coating knives. The thin film samples were then placed under vacuum for complete removal of the acetone. Absorption spectra were obtained under a nitrogen purge with an ATI Mattson Genesis Series FTIR utilizing $4000\text{-}600\text{cm}^{-1}$ wavelenghts.

Differential Scanning Calorimetry

Glass transition temperatures were obtained on a TA Instruments 2920 Differential Scanning Calorimeter equipped with a TA Instruments Liquid Nitrogen Cooling Accessory (LNCA). Samples were heated using a heat/cool/heat method at a rate of 3°C/min over a temperature range of −20°C to 100°C under a helium purge.

To ensure similar thermal history of each sample, glass transition temperatures were taken from the inflection point of the second heat curve.

Dynamic Mechanical Analysis

Samples were prepared by pressing dyes of 1.6mm thick, 9.8mm wide, and 25.2mm long between ferrotyping plates in a hot press at 200°C with a pressure of 5 metric tons. Dynamic mechanical data was obtained on a TA983 DMA at a rate of 3°C/min over a temperature range of –150°C to 100°C. Flexural mode was used with 0.4mm amplitude at 3 Hz.

Dielectric Analysis

Dielectric measurements were conducted on a TA Instruments DEA2970. Under a nitrogen purge of 500mL/min, ceramic parallel-plate sensors, screen- printed with gold, were used to measure dielectric properties of the blends. A sample with dimensions of 0.8mm thick and 25mm in diameter was subjected to an applied voltage which produced a permittivity and loss factor that were recorded as a function of frequency and temperature. A frequency sweep over a range of 0.1 to 300,000 Hz from –150°C to 200°C was conducted with a heating rate of 3°C/min. The ram applied a maximum force of 250N which produced minimum spacing of 0.5mm for each sample

Results and Discussion

UV/VIS

The miscibility of the copolymer blends was examined using transmission data. At 550nm a maximum 93% transmission can be seen for pure CP-41, and only 3% transmission can be seen for pure Kynar SL. The % transmission is not expected to reach 100% for CP-41 due to a reflection loss per surface of 3.4% calculated by [15]:

$$R = \frac{(n_1 - n)^2}{(n_1 + n)^2}$$

Increasing Kynar content in the blend diminishes transparency because of the mismatch in refractive indices between the methacrylate/Kynar miscible material and the phase separated Kynar portion. The area under the transmission curve was calculated for each of the blends by computing the integral of y with respect to x using the trapezoidal rule and is plotted in Figure 1. Films up to 20% Kynar are equally transparent. As evidenced by the area under the transmission curve, the data suggest the onset of phase separation at 20% Kynar. Prior UV studies have suggested the onset of phase separation for the homopolymer blends to begin at 40%, however % transmission was integrated only over the visible range [1].

Refractive Index and Numerical Aperture

Refractive index data is reported in Table 1. The refractive index of CP-41 copolymer is 1.497, and is near that of the homopolymer PMMA of 1.489. Blending the methacrylate copolymer with Kynar decreases the refractive index from 1.497 for the neat CP-41 to 1.471 for the 20% Kynar blend (the onset of phase separation via UV/VIS).

To assess the effectiveness of these coatings as fiber cladding materials, the numerical aperature, N_a, of these hypothetical fibers were calculated. N_a is a function of the refractive index of the core material, n_1, and cladding material, n_2 [16]:
$$N_a = (n_1^2 - n_2^2)^{0.5}$$

Numerical aperture is a measure of how well light propagates down the fiber core; a higher N_a value indicates greater light harvesting efficiency. Table 2 summarizes N_a results for fibers clad with the Kynar blends. Values are calculated using polystyrene core material ($n_2 = 1.59$) and a typical silica core material ($n_2 = 1.50$). With polystyrene core, blending the methacrylate copolymer with Kynar increased N_a from 0.536 for 0% Kynar to 0.604 for 20% Kynar, as seen in Table 2. With a silica core, blending the methacrylate copolymer with Kynar increased N_a from 0.096 for pure CP-41 to 0.296 for 20% Kynar.

FTIR

FTIR spectroscopy was used to determine miscibility since hydrogen bonding interactions produce the most observable effect in studies of blend miscibility [9]. Absorption data in Figure 2 exhibit both frequency shifts and changes in peak intensity of the spectra due to interactions between the Kynar and CP-41 on the molecular level. The frequency shifts of the carbonyl stretch at 1730cm^{-1}, the assymmetrical CH_3 bend at 1450 cm^{-1}, the symmetrical CH_3 bend at 1388 cm^{-1}, and the CH_2 twist are reported in Table 3. The change in these frequencies indicates mixing of the copolymers on the molecular level. Figure 2 also indicates an increase in peak intensity for increasing percent Kynar in the blend for the C-F peak at 840 cm^{-1}, becoming visible at 25% Kynar. The change in band intensity results from hydrogen bonding between the methacrylate and vinylidene difluoride chains [9,17]. This data matches that of the homopolymer blend of MMA/VDF [9].

The change in frequency for the CP-41 methylene group at 1388 is depicted in Figure 3 showing this effect. A linear increase is observed with increasing Kynar content with a change in slope at the onset of phase separation at 25% Kynar. Once

Table I. Refractive index data for Kynar/CP-41 blends.

Material (wt.%Kynar)	Refractive Index
0%	1.4969
10%	1.4760
15%	1.4715
20%	1.4706
25%	1.4642
30%	1.4584
35%	1.4581
100%	1.4100

Table II. Numerical aperture data for Kynar/CP-41 blends.

Material (wt.%Kynar)	NA (polystyrene)	NA (silica)
0%	0.536	0.096
10%	0.591	0.2672
15%	0.602	0.2910
20%	0.604	0.2955
25%	0.620	0.3258
30%	0.633	0.3508
35%	0.634	0.3521

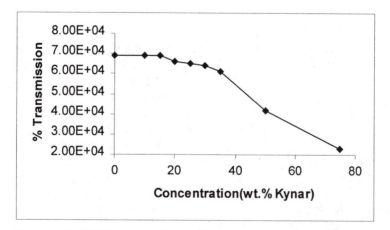

Figure 1. UV/VIS integrated transmission spectra from 190-820nm.

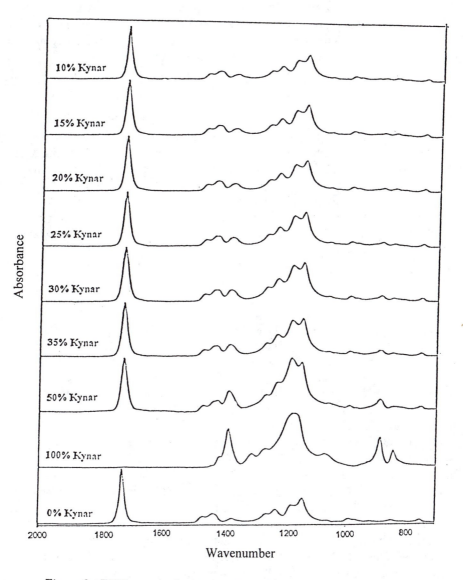

Figure 2. FTIR spectra for Kynar/CP-41 blends.

the Kynar begins to phase separate, hydrogen interactions only increase slightly with little dependence on Kynar content. The shift change relates a specific interaction of the CP-41 CH_3 with the Kynar. Some experiments in the past have used addition or subtraction methods to approximate compatibility [9,17]. We were unable to use this method in regions depicting the onset of phase separation. Coleman et.al reported difficulty approximating the spectrum in this region as well [16].

DSC

The glass transition temperatures for the blends are shown in Table 4. A single glass transition is observed for blends with Kynar content up to 75%. In addition to a glass transition, the immisible 50% Kynar and 75% Kynar blends also exhibited a melt temperature indicating phase separation. In Figure 4, a plot of glass transition vs concentration of Kynar reveals a linear correlation from 0-25% Kynar, and a break in this trend where the Kynar phase separates. This indication of phase separation is consistent with phase separation data exhibited from FTIR, as well as the following techniques.

DMA

Figures 5-9 depict DMA data the Kynar SL/CP-41 blends and neat polymers. The storage modulus (E') represents the elastic component, or energy storage, of the polymer. The loss modulus (E") represents the viscous component, or energy dissipation, of the polymer. Plotting E' as a function of temperature evidences the use limits for stiffness; while plotting E" as a function of temperature evidences transitions. Kynar has a plasticizing effect on PMMA [13]. Loss and storage modulus data demonstrate this trend.

Loss modulus is reported in Figure 5 for 100% Kynar. Transitions closely resemble those found in previous dynamic mechanical data for the pure homopolymer PVDF [4]. The transitions below −50°C are attributed to segmental motion in the amorphous phase. The transition at 40°C is attributed to motion in the crystalline region. The transition at 0°C is also attributed to the amorphous region as weak local motion. The transitions for the copolymer are at lower temperatures than those found for the pure PVDF due to the suppression of the T_g by the PTFE [18].

Loss modulus data for 0% - 35% Kynar is reported in Figure 6. The temperature interval studied was −150-100°C. The broad transition between −15 and 50°C corresponds to the beta transition, and is attributed to rotational movements of the side chain ester groups [4]. The glass transition above 50°C indicates large scale motion or chain slippage. Segmental motion associated with T_g is more affected by plasticization than the side group motion associated with T_β. The observed T_g decreases more rapidly than the beta transition. This results in decreased separation between the two transitions, as can be seen in Figure 7. However, as crystalline

Table III. Frequency shifts due to plasticizing effect of Kynar
on CP-41. (wt. % Kynar)

%Kynar	ν(C=O)	σ(CH₂) σ (CH₃-O)	σₛ(α-CH₃)	(CH₂)twist
0%	1734.0	1447	1386	1149
10%	1732.7	1447	1389	1150
15%	1732.44	1447	1390	1150
20%	1732.92	1447	1391	1151
25%	1732.1	1446	1394	1151
30%	1732.16	1440	1397	1152
35%	1731.45	1439	1397	1152
50%	1730.43	1436	1399	1155
100%	N/A	N/A	1401	1182

Table IV. DSC data derived from Kynar/CP-41 blends.

Sample (Wt. %Kynar)	T_g	T_m
PVDF/PTFE		
CP-41	85.4	130
10%	76.7	
15%	70.8	
20%	63.1	
25%	59.8	
30%	60.6	
35%	60.1	
50%	46.5	107.5
75%	45.2	107.5

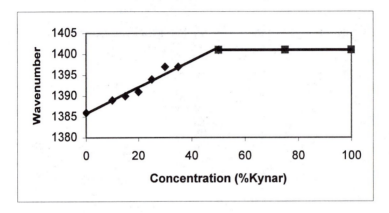

Figure 3. FTIR data representing concentration vs wavenumber.

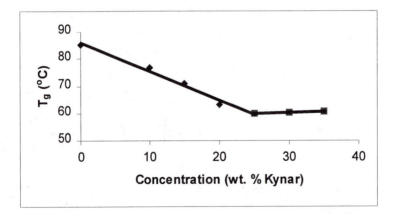

Figure 4. DSC plot of T_g vs % concentration of Kynar for 3 °C/min.

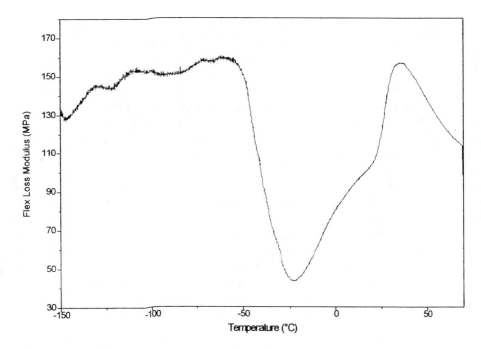

Figure 5. DMA loss modulus data for 100% Kynar.

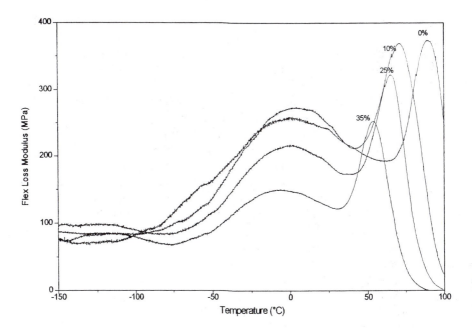

Figure 6. DMA loss modulus data for Kynar/CP-41 blends.

Kynar content is increased in the blend, both the T_g and beta transition temperatures decrease. This decrease demonstrates the plasticizing effect of the Kynar on the methacrylate copolymer. Also, the T_g and T_β transition move closer together with increasing Kynar content which is plotted in Figure 7.

Storage modulus is plotted in Figure 8 for 100% Kynar. A decrease in storage modulus with increasing temperature is observed. Storage modulus data for 0%-35% Kynar blends is represented in Figure 9. The E' data supports the evidence shown in E" for the plasticizing effect of the Kynar. Pure CP-41 exhibits a drop in modulus that is associated with the onset of the glass transition. A decrease in storage modulus is observed with increasing Kynar content. The temperature at which the modulus sharply drops also decreases with increasing Kynar. E' is much lower for the blend than for the pure amorphous polymer, indicating an increase in flexibility. All of this indicates that while the amorphous blends exhibit increased flexibility, their use is limited to temperature ranges below the onset of the glass transition.

For the copolymers studied, indications of phase separation are not evident from the DMA data, yet indications of compatibility are exhibited with E" transitions. Under conditions specified in this work, only one T_g can be detected for each blend, regardless of concentration. DMA transitions are better separated than those reported in upcoming DEA because conductivity effects mask the transitions in low frequency DEA results.

DEA

Dielectric analysis measures both permittivity (ε') and loss factor (ε") as a funcion of temperature. Figure 10 depicts the loss factor vs temperature for 100% Kynar at frequencies from 300MHz to 0.1Hz. Two separate relaxations are observed, an α_a corresponding to local motion in the glassy state and an α_c corresponding to the onset of melting. The PVDF/PTFE copolymer exhibits relaxations at higher temperatures compared to the relaxations of the pure PVDF. Also, the loss factor at frequencies below 10kHz and above 35°C suggest the copolymer is more conductive than the homopolymer as reported in previous work [12].

Each of the blends was studied over a frequency range of 0.1-300,000 Hz. The CP-41 copolymer of PMMA/PEA exhibits two transitions depicted in Figure 11. The β relaxation at lower temperatures is attributed to the rotation of the side chain when large scale motion is frozen. The $\alpha\beta$ transition corresponds to the T_g [10] which increases with increasing frequency. The miscible copolymer blends of 15% and 25% Kynar exhibit similar transitions to those of CP-41, as seen in Figure 12a and b respectively. A decrease in temperature for the $\alpha\beta$ relaxation is observed, while the β relaxation only slightly decreases with increasing Kynar content. As reported in Table 5, the data for the $\alpha\beta$ transition was fit to the Williams, Landel, and, Ferry (WLF) equation [20]:

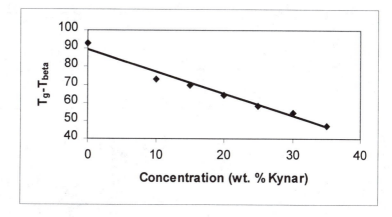

Figure 7. DMA data for Kynar/SLCP-41 blends.

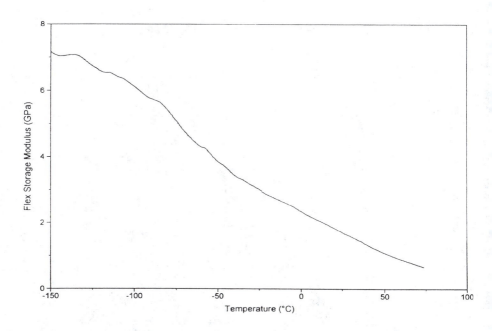

Figure 8. Storage modulus data for 100% Kynar SL.

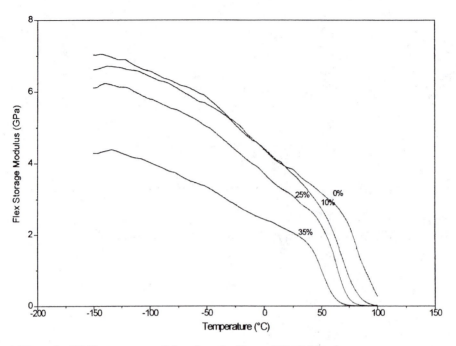

Figure9. DMA storage modulus data for Kynar/CP-41 blends.

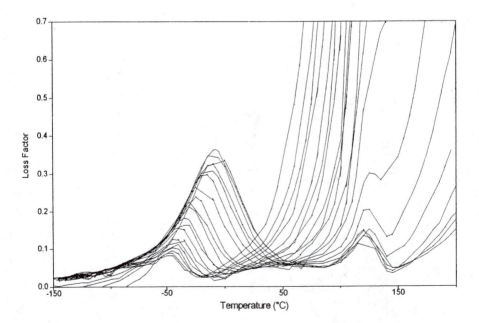

Figure 10. DEA relaxations for 100% Kynar.

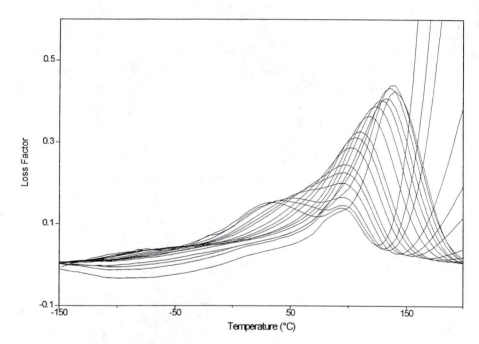

Figure 11. Transitions observed for pure CP-41 by DEA.

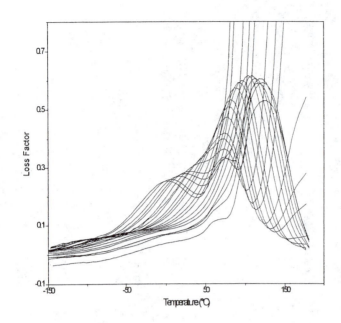

Figure 12a. Relaxations in DEA for 15% by wt. Kynar/CP-41 blend.

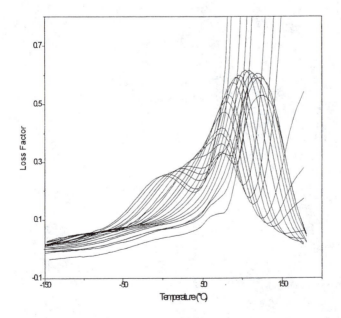

Figure 12b. DEA relaxations for 25% Kynar/CP-41 blend.

$$\log \alpha_t = \frac{-C_1 (T - T_0)}{C_2 + (T - T_0)}$$

where α_t corresponds to frequency, C_1 and C_2 are WLF constants, T is a given temperature , and T_0 is the reference temperature which corresponds to T_g. The WLF constants vary from the 'universal' constants $C_1 = 17.44$ and $C_2 = 51.6$. Such variations have been explained by differences in cooling rate, the presence of residual monomer, and moisture [20]. Activation energies were calculated using the Catsiff and Tobolsky equation [21]:

$$\Delta H = 2.303 \, (C_1 / C_2) RT_g^2$$

This equation uses the WLF constants to provide activation energies for each of the copolymer blends. These constants also provide information on free-volume in the glass transition region (f_g) and the thermal expansion coefficient (α_f) which are also shown in Table 5. The free volume and the thermal expansion coefficient decrease until the onset of phase separation at 25% Kynar. This behavior indicates that polymer chains pack more efficiently due to increased H-bonding interactions, observable in the FTIR data. As expected, the WLF behavior reveals a decrease in activation energy with increasing Kynar content, since the β relaxation contributes less to the $\alpha\beta$ process. The data begins deviating from WLF behavior at 25% Kynar and loses fit above this concentration due to the increasing crystalline content of the blend [22]. Figure 13 reports WLF curve fit for the 10% Kynar blend.

The β transition for the blends closely follows Arrhenius behavior as depicted in Figure 14 for the 25% Kynar blend. The activation energies for this relaxation are calculated from Arrhenius plots of ln frequency versus 1/T(K) and are reported in Table 6a. The activation energy increases as Kynar content increases due to increased molecular mobility in the amorphous regions. After the onset of phase separation, a transition is seen at much lower temperatures. For example, at 100 Hz the transition for the 75% Kynar blend is around -49°C compared to the transition for the 25% blend which occurs at 40°C. This lower temperature transition is no longer due to the amorphous portion of the blend, it is now due to the α_c transition in the crystalline portion. Table 6b reports the activation energy also calculated from Arrhenius plots of the lower temperature α_c for the phase separated blends.

Figure 15 represents ε" vs temperature for 20%-75% Kynar blends at 100Hz. Three transitions are seen. The $\alpha\beta$ transition which corresponds to the T_g for the CP-41, is seen around 75°C in the 20% and 25% blends. The transitions in these two blends fit WLF data for plasticized PMMA; the WLF fits, again, are in Table 5. At the onset of phase separation at 25-30% Kynar, a slight shoulder appears which is attributed to the presence of the α_c merging with the $\alpha\beta$ relaxation. However, the merged process is not as apparent at lower concentrations as it is for 50% and 75% Kynar where a broad relaxation is observed. Also, a second transition emerges at

Table 5. WLF behavior for $\alpha\beta$ relaxations as weight % Kynar.

	00	10	15	20	25
ΔH	96.6	89.6	78.8	78.8	78.7
T_g	73.28	59.69	56.9	54.7	54.43
C_1	10.4	10.4	11.6	11.8	11.2
C_2	59.1	58.8	73.3	73.6	69.8
f_g	0.041	0.041	0.037	0.037	0.039
α_f	7.07	7.11	5.10	5.00	5.56

Table 6a. Arrhenius behavior from β relaxations as weight % Kynar.

%Kynar	E_a
0	16.24
15	15.95
20	15.99
25	18.27
30	18.51
35	18.55

Table 6b. Arrenhius behavior obtained from α_c relaxation as wt. % Kynar.

% Kynar	E_a
50	20.74
75	18.39
100	23.40

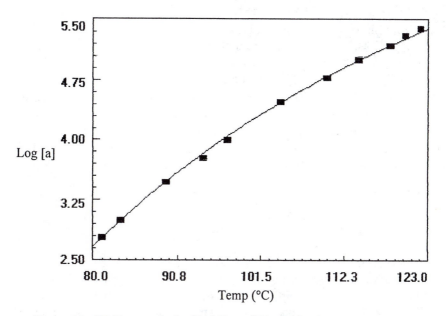

Figure 13. WLF curve fit for 10% Kynar/CP-41 blend.

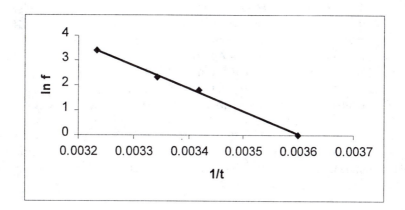

Figure 14. Arrhenius behavior followed by β relaxation for 25% Kynar blends.

lower temperatures for α_a corresponding to local motion in the glassy state of Kynar for blends of 35% to 75% Kynar [12].

When the temperature of the $\alpha\beta$ transition is plotted against the concentration of Kynar at 100 Hz, a change in T_g is illustrated beginning at 25% Kynar. Figure 16 displays this change with a variation in slope at 25% and again at 35% indicating the onset of phase separation. This plot proves this transition is due mostly to the $\alpha\beta$ relaxation, not the α_c relaxation, because above 35% Kynar T_g remains constant.

Prior to phase separation, the dielectric analysis of the copolymer blends exhibited relaxations similar to those of pure CP-41. The emergence of α_c at 30% Kynar implies the onset of phase separation. The presence of the α_a relaxation for blends of 35% Kynar and greater, also implying that two separate phases exist.

Conclusion

Several techniques were used to investigate the miscibility of the copolymer blends of CP-41 with Kynar SL. From UV/VIS data, the copolymer blends exhibited a transparent phase up to 20% Kynar. Subsequent analysis conducted using FTIR, DSC, and DEA show the onset of phase separation at 25% Kynar. The differences in the identified solubility range may be ascribed to variations in instrumental sensitivity. FTIR shows frequency shifts and changes in band intensity beginning at 25% Kynar relating interactions on the molecular level. DSC data exhibits a leveling off of T_g at 25% Kynar and at 50% Kynar, showing a change in phase behavior. Dielectric data shows the emergence of the α_c transition for the crystalline portion of the 25% Kynar blend as well as the emergence of the α_a transition at lower temperatures. In addition, the DEA data starts deviating from WLF behavior at 25% Kynar. The plasticizing effect of the Kynar in CP-41 is evident from the DMA data, as observed from decreased glass transition temperatures.

The copolymer blends characterized in this study exhibited a single, transparent phase at Kynar contents of 20- 25%. This study is significant in that it reveals insight on phase behavior and the onset of phase separation in fluorocarbon methacrylate copolymer blends. These blends may be used as fiber cladding materials, coatings for planar and rectangular wave-guides, and in optical applications requiring low refractive index transparent coatings.

Acknowledgments

This work was supported by the University of South Florida, the State of Florida, and Honeywell, Inc. together via the I-4 Corridor Initiative.

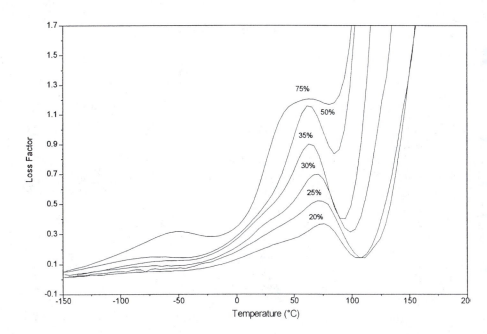

Figure 15. Loss factor vs temperature for 20% through 75% Kynar blends at 100 Hz observed by DEA.

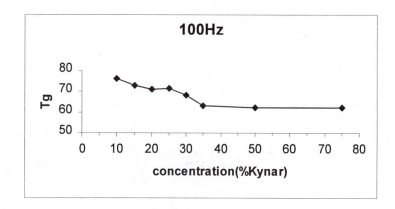

Figure 16. DEA data for αβ relaxation at 100Hz.

References

1. Noland, J.S., N. N.-C. Hsu, R. Saxon, and J.M. Schmitt, *Adv. Chem. Ser.* **1971**, *99*, 15.
2. Nguyen, T., *Rev. Macromol. Chem. Phys.*, **1985**, *C(25)2*, 227-275.
3. Bernstein, R.E., D.C. Warmund, J.W. Barlow, D.R. Paul, *Poly. Eng. and Sci.* **1978**, *18*, 1220.
4. Linares, J.L. Acosta, *Poly. Bull.* **1996**, *36*, 241.
5. Garcia, J.L., K.W. Koelling and R.R. Seghi, *Poly.* **1997**, *32*, 3861.
6. Gaynor, J., V. Fischer, J.K. Walker, J.P. Harmon, *Nuc. Instr. and Meth. in Phy. Res.* **1992**, *B69*, 332.
7. Jouannet, D., T.N. Pham, S. Pimbert, and G. Levesque, *Poly.* **1997**, *38*, 5137.
8. Cypcar, C. C., L. Jodovits, *Poly. Pre.*, **1998**, *39*, 861.
9. Koenig, J. L., *Spec. of Poly.*, Am. Chem. Soc., Washington D.C. **1992**.
10. Aihara, T., H. Saito, T. Inoue, H-P. Wolff, and B. Stuhn, *Poly.* **1997**, *39*, 129.
11. Walsh, J., .D.J., Higgins, J.S., *Polymer Blends and Mixtures*, Martinus Nijhoff Publishers, Dordrecht **1985**.
12. Mijovic, J., J. Wing Sy, and T.K. Kwei, *Macromolecules*, **1997**, *30*, 3042.
13. Martinez-Salazar, J.C. Canalda Camara, F.J. Balta Calleja, **1990**.
14. Yano, S., *J. of Poly. Sci. Part A*, **1970**, *8*, 1057.
15. *Engineering Plastics, Vol. 2 Engineered Materials Handbook*, ASM International, Metals Park, OH **1988**.
16. Miller and Chenowyth, *Optical Fiber Telecommunication*, Academic Press, New York **1979**.
17. Coleman, M.M., Zarian, J., Varnell, D.F., Painter, P.C., *Poly. Lett. Ed.*, **1977**, *15*, 745-750.
18. Menard, K. P., *Dynamic Mechanical Analysis*, CRC Press, Boca Raton, **1999**.
19. Aklonis, J., MacKnight, W., Shen, M., *Introduction to Polymer Viscoelasticity*, Wiley Interscience, New York **1972**.
20. McCrum, N. G., Read, B.E., Williams, G., *Anelastic and Dielctric Effects in Polymeric Systems*, John Wiley and Sons, New York, **1967**.
21. Catsiff, E., Tobolsky, A.V., *J. of Colloid Sci.*, **1955**, *10*, 375.
22. Van Krevelin, D. W., *Properties of Polymers*, Elsevier, New York, **1976**.

Chapter 8

Fluorinated Polymer Blends as Plastic Optical Fibers Cladding Materials

S. Pimbert[1], L. Avignon-Poquillon, and G. Levesque

Laboratoire Polymeres et Procedes, Université de Bretagne-Sud, Rue St. Maudé, 56325 Lorient, France
[1]Corresponding author: email: sylvie.Pimbert@univ-ubs.fr

Fluorinated polymers and copolymers blends have been studied as potential cladding materials for plastic optical fibers (POF),. Vinylidene fluoride (VDF) homo and copolymers with hexafluoropropene (HFP) or trifluoroethylene (TrFE) were mixed with PMMA or MMA-trifluorethyl methacrylate (MATRIFE) copolymers. Miscibility domains in binary blends are more important when VDF copolymers are used as their crystallinity fraction is lower. The introduction MMA-MATRIFE copolymers to promote lower blend refractive indices gives reduced miscibility domains as compared with PMMA containing mixtures. Competition between intramolecular and intermolecular interactions is postulated to explain these results, also observed in ternary blends.

PMMA-core POF cladding with several blends was realized and gave good results, although the less crystalline PVDF-HFP or PVDF-TrFE copolymers might be used without blending. The phase nature and morphology near the interface in such fibers is questionable and we report preliminary crystallization kinetics results for some of these blends which may help to describe such interfaces.

Polymer optical fibers (POF) are well known for short distance light transmission and largely used for decorative displays or data transmission systems [1,2]. In a coextrusion process, POF are obtained by simultaneous extrusion of both core and cladding materials. When PMMA (T_g = 115 °C, n_1 = 1.492 (reference PMMA OF104S from Rhöm)) is choosen as core material, cladding material needs to fit requirements such as

- Low refractive index ($n_2 < 1.44$) to obtain high enough POF's numerical aperture.
- Sufficiently high Tg (above 60°C or 333 K) for a thermal behavior similar to PMMA (low temperature applications).
- Amorphous or low crystallinity material to avoid light diffusion at the interface region
- Excellent adhesion on PMMA.
- Low cost is of course preferable.

So looking for blends of commercial or readily available materials seemed to be a less expansive solution.

The « best » cladding material must present sufficient compatibility with PMMA to adhere strongly on it : mechanical stresses concentrate in the interface region during the fiber extrusion, stretching and cooling steps. Poor adhesion may induce materials separation during fibers lifetime, thus modifying their optical and mechanical performances.

In the preform POF technology, the core material is usually covered with a viscous solution of a copolymer solution often containing one or two monomers related to the copolymer cladding. Drying is then associated with a final copolymerization step and the resulting interface region is probably wider in such a process.

By contrast, coextrusion does not involve any contact between the core material and a solvent-monomer mixture able to swell it superficially. In such a process, core-clad interface is nearly equivalent to welding surfaces observed in injection molded parts. Diffusion or interpenetrating process does not have sufficient time to occur. Consequently, strong adhesion between co-extruded core and cladding materials might exist only with materials normally offering attractive specific interactions, *i.e,.* resulting in a miscibility domain in their phase diagram. For example, experimental data obtained on fibers from preform heating and stretching have demonstrated that PMMA and MMA-fluoro-alkyl methacrylate copolymers are not really miscible when the fluoro-alkyl content becomes too high : thus, MMA-trifluoroethyl methacrylate (MATRIFE) copolymers are not suitable for cladding when their MATRIFE content becomes higher than 80-84 % [3]. Nevertherless, poly(fluoroacrylates) having rather short fluoroalkyl side groups have been often copolymerized with other acrylic ester monomers and these copolymers were suggested for use as cladding materials [4].

POFs are becoming more and more used with the continuous co-extrusion process looks as the most suitable production process for step-index POF. Then, materials to be used as cladding on a PMMA core should be preferably extrudable (co)polymers, miscible with PMMA, at least partially.

Miscibility of PMMA with many halogenated polymers is well documented [5-7]. As fluorine atoms are desirable to lower the refractive index, PMMA-PVDF system was an excellent base to look for cladding materials from polymer blends.

Several blends similar to the well known PMMA-PVDF partially miscible system [8, 11] have been studied. They contain a methacrylic ester (homo- or co-) polymer and one or two fluorinated vinylic (homo- or co-) polymers. The presence of difluoro methylene groups in the vinylic polymers structural units induces strong electro-attractive effects on vicinal methylene C-H bonds. So, the miscibility of these systems is based on the existence of specific interactions between these polarized C-H and C-F bonds in the polyvinylic structural units and the carbonyl groups present in the polymethacrylic ester units.

In fluorinated vinylic copolymers, substitution of hydrogen atoms by species such as F atoms or CF_3 groups results in crystallinity loss and might induce an enlargment of miscibility domains. On the other hand, substitution of methyl ester groups in MMA copolymers by fluorine containing alkyl chain contributes to lower refractive indices [12].

We have mainly studied blends containing PVDF or its copolymers with hexafluoropropene (VDF-HFP) or trifluoroethylene (VDF-TrFE) on one side and PMMA and its copolymers with trifluoroethylmethacrylate (MAT 33 and MAT 66) on the other side.

Results on the miscibility of such binary and ternary blends are first presented. Then we will discuss about the opportunity to use some of these systems as POF cladding materials. We conclude with a short section on the crystallization behaviour of some blends.

Experimental Section

Materials

Atactic PMMA was kindly supplied by Röhm (PMMA OF 104S) and fluorinated vinylic polymers : polyvinylidene fluoride homopolymer (PVDF), vinylidene difluoride-hexafluoropropene (VDF-HFP) copolymer and vinylidene fluoride-trifluoroethylene (VDF-TrFE) copolymer were obtained from Solvay S.A. and Elf Atochem. Side-chain fluorinated methacrylic homo- and copolymers were synthesized by free-radical suspension polymerization.

Monomers were purified by filtration through an activated basic alumina column, then saturated with dry nitrogen during one hour and stored under N_2 at 5°C. Vacuum-distilled dodecanethiol (DDT) was used as chain tranfer agent and azobis*iso*butyronitrile (AIBN), recrystallized in ethanol, as polymerization initiator. Aqueous phase made of distilled water (3.5 liters), polyacrylic acid (5 grams) and Na_2HPO_4 (45 grams) was placed into a 5-liter Pyrex reactor at the polymerization temperature (between 55 and 65 °C according to the composition) and saturated with nitrogen for 2 hours. Then, under stirring, monomers (1500 grams), AIBN (4.5 grams) and DDT (4.5 grams) were rapidly introduced. Polymerization progress was followed by temperature evolution. An exothermic period was observed after five to six hours. The reactor was still kept at 55 °C for 10 hours then heated to 70°C for 10 hours. Polymers beads were separated onto a sintered glass filter and washed several times with distilled water. After drying under vacuum at 55-60 °C for 48 hours. *ca.* 1 millimeter diameter polymer beads were obtained.

The main characteristics and chemical structures of these polymers are described respectively in Table 1 and Figure 1.

Analysis for remaining monomer indicated less than 0.2% volatile materials.

Table I. Main characteristics of homo and copolymers used in blends

Polymer	Mn (a) g/mol	Mw (a) g/mol	$T_g(K)$	$T_m(K)$	$-\Delta H_m$ (J/g)	n_D
PVDF	39 000	110 000	245.0	448.1	66	1.4200
VDF/HFP 90/10	43 000	120 000	250.8	434.7	30	1.41 (c)
VDF/HFP 80/20	50 000	120 000	259.5	408.9	22	1.41 (c)
PTrFE (b)	nd	nd	nd	470.7	22	1.3890
VDF/TrFE 75/25 (b)	nd	nd	nd	424.0	34	1.4125
VDF/TrFE 50/50 (b)	nd	nd	nd	428.0	26	1.4050
PMMA	57 510	123 400	391.0	-	-	1.4920
PMATRIFE	60 300	108 200	346.0	-	-	1.4200
MATRIFE/MMA 33.3/66.6	59 400	98 400	379.4	-	-	1.4679
MATRIFE/MMA 66.6/33.3	75 600	118 000	370.0	-	-	1.4440

(a) Size Exclusion Chromatography, PMMA standards, THF

nd : non determined

(b) MFI 230°C-2.16 Kg

PTrFE : 2.7 cm³/10mn   VDF-TrFE 75-25 : 1.6 cm³/10mn   VDF-TrFE 50-50 : 2.3 cm³/10mn

(c) Supplier data

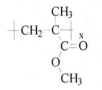

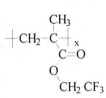

PMMA

PMATRIFE

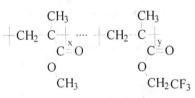

MMA – MATRIFE

PVDF

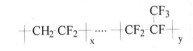

VDF – TrFE

VDF - HFP

Figure 1 : Fluorinated and methacrylic copolymers used

Blends were prepared at 20-50 °C by solution mixing in dimethylformamide, as 5% (w/w) solutions, followed by fast precipitation in distilled water. Blends were then filtered off, washed several times with water to eliminate any solvent trace and dried *in vacuo* at 70°C to constant weight.

For refractive indices measurements, some blends were prepared from the melt using an opened two-rolls mixer at 210°C for 10 mn. These mixtures were then pressed under 200 bars at 210°C, cooled and cut into small plates.

Differential Scanning Calorimetry (DSC). A Perkin-Elmer DSC 7 apparatus equiped with an Intracooler II cooling (for binary blends) and a Perkin Elmer PYRIS 1 equipped with a Cryofill system (for ternary blends) were used to determine transition temperatures, using indium and tin calibration samples, under nitrogen atmosphere.

All samples were submitted to the same temperature program : 5 min isotherm at 298 K, heating from 298 K to 483 K at 20 $K.mn^{-1}$, 5 min isotherm at 483 K, quenching to 218K at 200 $K.mn^{-1}$, 5 min isotherm at 218 K, second heating to 483 K at 20 $K.mn^{-1}$. Glass temperatures were recorded at the half height of heat capacity jumps. Melting temperatures were considered at endothermic peaks maxima.

For the crystallization study, a slighty different temperature program was used : : 5 min isotherm at 298 K, heating from 298 K to 483 K at 20 $K.mn^{-1}$, 5 min isotherm at 483 K, cooling to 298K at 10 $K.mn^{-1}$, 5 min isotherm at 298 K, second heating to 483 K at 10 $K.mn^{-1}$

Refractive index. A RL-1 refractometer from PZO (Warzaw, Poland) was used to determine the refractive indices of small plates of the polymers at 23°C.

Results and Discussion

Binary blends from solution mixing

PMMA-containing blends obtained by substituting PVDF by its copolymers VDF-HFP 90/10 and 80/20 (w/w) have been studied. DSC results were compared with the PMMA-PVDF system. In figure 2, we observe complete miscibility with a single Tg up to 50% (w/w) VDF-HFP (90/10) content and 70% (w/w) VDF-HFP (80/20) content, whereas miscibility is observed only up to 40 % (w/w) PVDF content. For higher contents, these blends become partially crystalline. In the crystalline domains, we can also notice a decrease of the melting temperatures in fluorinated copolymer blends as compared to PVDF system. These enlargments of miscibility domains may be justified by the lower melting enthalpies in VDF copolymers (22 $J.g^{-1}$ for VDF-HFP (80/20) and 30 $J.g^{-1}$ for VDF-HFP (90/10) compared to 66 $J.g^{-1}$ for PVDF.

For all blends, Tg regularly decreases with the vinylic polymer content in the amorphous phase until VDF-HFP copolymers crystallisation occurs.

Our study was extended to blends of VDF-HFP 80/20 copolymers with a fluorinated methacrylic copolymer obtained by substitution of PMMA by MMA-trifluoroethyl methacrylate (MATRIFE) copolymers, designed as MAT33 and

MAT66 (respectively 33% and 66 % w/w MATRIFE/MMA copolymers). Large miscibility domains are observed in figure 3. For all blends we note only one Tg in the amorphous domains ; the Tg values are strongly correlated to the blend composition.

Introduction of MATRIFE units in methacrylic ester copolymer leads to a decrease of the glass transition temperature ; miscibility domains widths are also reduced.

Blends are miscible until 70% (w/w) vinylic fluorinated copolymer in PMMA/VDF-HFP (80/20) system whereas the limit comes back to 60% (w/w) in MAT33/VDF-HFP (80/20) system and 50% (w/w) in MAT66/VDF-HFP (80/20) system. For higher vinylic polymer contents, a melting endotherm appears corresponding to partial crystallization of these blends.

Our results give evidence of the influence of fluoroalkyl lateral chains on the miscibility decrease in these systems. The highest MATRIFE unit content in copolymers (66% w/w) gives the less miscible blend. Although the introduction of fluoro-alkyl ester in MMA-copolymers would have been thought to improve their miscibility with PVDF and its copolymers, the contrary is observed and we refer to competition between *inter-molecular* and *intra-molecular* attractive interactions to explain such blend behaviours [13]. The presence of fluorinated methacrylic ester in MMA-copolymers necessarily introduces new but *intra-molecular* dipolar interactions, unfavorable to miscibility for entropic considerations. Because of the high fluorine atom electronegativity, the vicinal hydrogen of methylene groups in fluoroalkyl ester groups or fluorinated vinylic chain become more positively charged through C-H bond polarization. Therefore, these hydrogen atoms become concurrent to create *inter-* as well as *intra-molecular* interactions with carbonyl groups present in methacrylic units.

Blends of PMMA with an other fluorinated copolymer, VDF-TrFE copolymer, were studied to confirm this hypothesis. As for HFP monomers, TrFE units introduce irregularities in the PVDF chains and lead to less crystalline copolymers with lower melting temperatures. Moreover, PTrFE homopolymer presents a lower refractive index than PVDF (1.3890 against 1.4200). Thus it could be interesting to use blends of VDF-TrFE copolymers as cladding materials. DSC results are presented in Figure 4 for PMMA-(VDF-TrFE) copolymer - blends. Only one glass transition is observed up to 70% (w/w) VDF-TrFE (75/25) content and up to 60 % (w/w) VDF-TrFE (50/50) content. For higher vinylic polymer contents, the blends become partially crystalline with melting temperature increasing slighty with the vinylic fluorinated copolymer mass fraction. The enlargment of miscibility domains for these blends in respect to PVDF-containing mixtures is also attributed to the lower crystallinity in VDF-TrFE copolymers compared to PVDF homopolymer. We did not consider the effect on the Curie transition of VDF-TrFE copolymers and blends because of the field of application not concerned by magnetic properties.

In the same way, the influence of methacrylic ester copolymer type on VDF-TrFE copolymers crystallization has been studied. As previously observed with VDF-HFP copolymers, the presence of fluoroalkyl ester groups in MMA-copolymers induces a narrowing of miscibility domains compared with those obtained with PMMA alone.

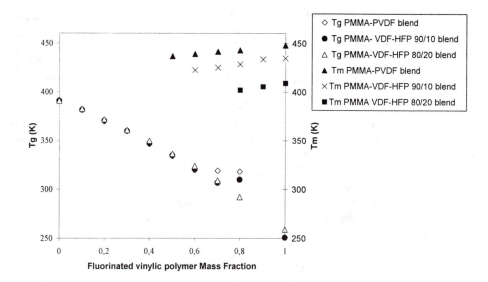

Figure 2 : DSC results on PMMA / PVDF and PMMA / (VDF-HFP) blends

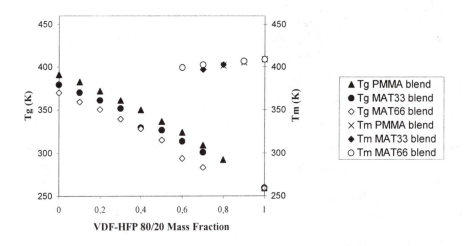

Figure 3 : DSC results on (MATRIFE-MMA) – (VDF-HFP 80/20) copolymer blends

Ternary blends from solution mixing

Our study was enlarged to ternary blends to get more flexibility in the material properties : a third constituant was added to binary systems previously characterized. Miscibility domains were compared.

Studies of ternary systems reported in the literature [14-17] deal with the addition to two immiscible polymers of a third polymer miscible with each of the component of the binary system to produce a fully miscible ternary blend. Polymers choosen for binary systems are generally amorphous. For example, results of the addition of polyvinylchloride (PVC) in a system of two immiscible polymethacrylates have been described (13). Blends containing 30% or more of PVC show a single glass transition temperature. PVDF was also used in association with poly(methylmethacrylate) and poly(ethylmethacrylate) (14). Addition of polyvinylacetate (PVAc) to blends of PVDF and PS has also been studied (15). So it seems interesting to investigate the possibility to obtain miscible ternary blends by adding a methacrylic homo or copolymer to an immiscible system of two fluorinated semi-crystalline copolymers.

First, the lack of miscibility between PVDF and its fluorinated copolymers was characterized. DSC results show that, despite their very similar chemical structure, such binary blends are totally immiscible. Whatever the system composition, two melting temperatures are observed, nearly identical to melting temperatures of the pure components.

The influence of introducing a methacrylic ester copolymer (PMMA, MAT33 or MAT66) in an immiscible binary system of two fluorinated vinylic copolymers, respectively VDF-HFP 80/20 and VDF-TrFE 50/50, was studied. Results are presented in Figure 5. The ternary phase diagram exhibits a large miscibility domain for blends having a sufficient methacrylic ester copolymer content. A single Tg is recorded. It is possible to obtain amorphous ternary blends : for methacrylic ester copolymer content between 35 and 55 %, it depends on the proportion of MATRIFE units in the acrylic copolymer. For less than *ca.* 25% PMMA, all blends are immiscible with the presence of one or two melting endotherms corresponding nearly to the melting of pure components. The largest miscibility domain is obtained with PMMA. We observe a narrowing of the amorphous domain when MATRIFE units are introduced in copolymers with MMA. These results afford confirmation of the previous experiments concerning the influence of fluoroalkyl ester chain on the miscibility decrease in these systems. The introduction of *intra-molecular* interactions in competition with *inter-molecular* interactions between the carbonyl groups and the acidic hydrogen contribute to the miscibility loss occuring in such systems.

Similar results were obtained with the ternary system : PVDF / VDF-TrFE(50/50) / methacrylic ester copolymer (PMMA, MAT33 or MAT66) (Figure 6). Substitution of VDF-HFP 80/20 copolymer by PVDF homopolymer - which is highly crystalline - induces a strong reduction of the miscibility domains. We confirm thus that the use of less crystalline copolymers is favorable to obtain miscible blends. As in ternary blends with VDF-HFP 80/20, the influence of the MATRIFE units content in methacrylic ester copolymer is obvious. Ternary phase diagram presented in Figure 6 shows a very

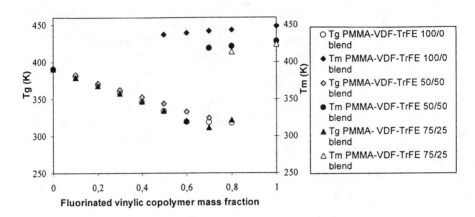

Figure 4: DSC results on PMMA–(VDF-TrFE) copolymer blends

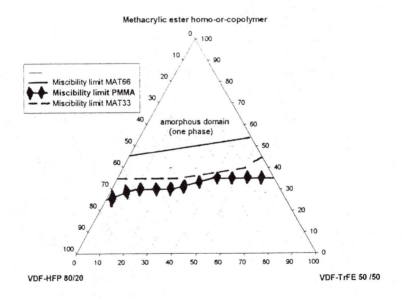

Figure 5 : (VDF-HFP 80/20) / (VDF-TrFE 50/50) / Methacrylic ester copolymer Ternary phase diagram

narrow miscibility domain with MAT66 copolymer. An enlargment of this domain is observed when using MAT33 and again with PMMA. When the percentage of MATRIFE units decreases, inter-molecular interactions between methacrylic ester copolymer and fluorinated copolymers become predominant with respect to intra-molecular interactions.

Applications in POF

As it appears in Figure 7, some blends may present sufficiently high Tg (>333K or 60°C) and low refractive index (n_D < 1.44) to be used as POF cladding materials, particularly in the case of MAT66/VDF-TrFE blends. Mixtures of these copolymers containing at least between 20 and 40 % VDF-TrFE present physical characteristics which fit the above requirements. They were tested with success as POF cladding in a co-extrusion process. In this type of process, strong adhesion is given as a result of materials miscibility at the interface.

However, the good miscibility between PMMA and the less crystalline VDF copolymers also allows the use of these materials without blending as cladding on PMMA, although the core material is directly in contact with a semi-crystalline polymer. A mixed amorphous interface zone might have been formed which should avoid light diffusion from crystals in the pure cladding materials. However, the melt/mixing process between macromolecules is sufficiently low to limit the true depth of such an interface. It seems more likely to postulate that near the interface, the crystallization kinetics is sufficiently modified to induce crystals dimensions below the light diffusion limit.

In order to get more data concerning the physical state of the cladding polymer in the interface region between PMMA and VDF-copolymers, crystallization experiments were conducted to investigate the influence of both time and temperature on the phase type and morphology in some blends. Preliminary results are presented below.

Crystallization Behaviour in Fluorinated Copolymers Blends

We have focused our work on some binary and ternary blends of PMMA, or MATRIFE-MMA copolymer, with PVDF, or one of its copolymers, VDF-TrFE, in the heterogeneous zone. The influence of PMMA and MAT66 on the crystallization behavior of VDF-TrFE (50-50) copolymer or of PVDF and VDF-TrFE copolymer (50-50) immiscible blend will be published into detail elsewhere.

First, a particular attention was given to PMMA-VDF-TrFE (50-50) binary blend in the range 0-40 % PMMA. The crystallization behaviour of this system was compared to that of MAT 66-VDF-TrFE (50-50) system, up to 55% MAT 66.

VDF-TrFE melting enthalpies (on second heating) have been determined by DSC for these blends and reported on Figure 8 versus methacrylic ester copolymer content. We may notice on this graph a regular decrease of the melting enthalpy per gram of

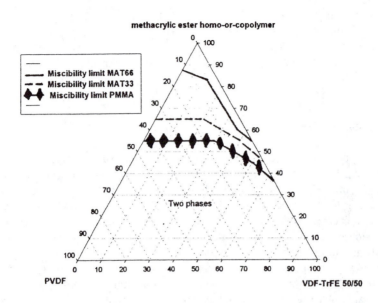

Figure 6 : PVDF / (VDF-TrFE 50/50) / Methacrylic ester copolymer Ternary phase diagram

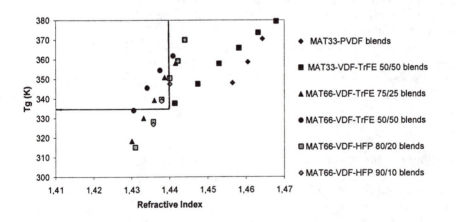

Figure 7 : Correlation between blends refractive indices and the glass transition temperatures

Lines represent the limits for materials to be used as POF cladding
(T_g>333 K and n_D<1,44)

semi-crystalline polymer used when the methacrylic ester copolymer content is increased; this corresponds to a decrease of the fluorinated copolymer crystallinity associated to a partial dissolution of the semi-crystalline copolymer in PMMA or MAT 66. VDF-TrFE solubility is lower in MAT 66 than in PMMA, as it appears on this graph. This can be justified by the existence of *intra-molecular* interactions between MMA and MATRIFE units in MAT 66 copolymer, in competition with *intermolecular* interactions between MMA and VDF units, as demonstrated above. The predominant character of *intra-molecular* on *inter-molecular* interactions contributes to slow down VDF-TrFE dissolution in the MATRIFE-MMA copolymer amorphous phase.

This crystallization study was extended to ternary blends with equal contents of the two fluorinated copolymers and a PMMA weight percentage varying from 0 to 50 %, which allows a direct competition between the two fluorinated vinylic polymers. Anisothermal crystallization and melting temperatures of PVDF and VDF-TrFE (50-50) copolymer, as well as the corresponding enthalpies were determined by DSC. As it appears on Figures 9 and 10, we may observe a decrease of both T_c and T_m when the PMMA content is increased from 0 to 30 % (w/w). Only one crystallization exotherm is observed justified by very near values of T_c for these two polymers (respectively 139°C and 137°C for pure PVDF and VDF-TrFE 50-50). At 42 % PMMA, the blend is metastable : crystallization occurs only on heating above T_g.

The evolution of melting enthalpies versus PMMA content in these ternary blends is reported on Figure 11. The measured melting enthalpies are expressed in Joule per gram of fluorinated polymer and then corrected for each semi-crystalline copolymer content in the ternary blend. A decrease of ΔH_m for the VDF-TrFE copolymer fraction is observed until 30 % PMMA is reached in the blend, whereas the PVDF fraction melting enthalpy is nearly constant. When all VDF-TrFE is dissolved in PMMA to form an amorphous phase, PVDF begins to dissolve in this amorphous blend; this occurs for PMMA contents between 30 and 42 % and is expressed by a decrease of PVDF melting enthalpy. Beyond 42% PMMA, this ternary blend is completely amorphous : no melting peak is observed indicating a totally miscible system. These results may be correlated with the crystallinity differences of the two fluorinated polymers : the less crystalline copolymer is dissolved preferentially before the VDF homopolymer. After copolymer has been completely incorporated in the amorphous phase, PVDF may in turn begin to be dissolved.

Crystallization data in binary and ternary blends containing PMMA or MATRIFE-MMA copolymers associated with VDF-copolymers appear to be strongly correlated with the crystallinity ratio in this last polymer. Further experiments are devoted to crystallization kinetics and morphological studies of the crystals obtained in such mixtures. The influence of many other parameters need to be examined, such as temperature gradients, molecular weights and distribution, chemical bonds at the interface, to achieve a better knowledge of the interfaces between PMMA and VDF-copolymers.

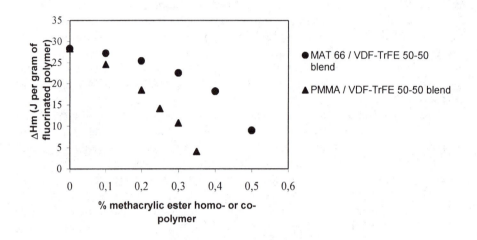

Figure 8 : Melting enthalpies in (PMMA or MAT 66)–(VDF-TrFE) copolymer blends

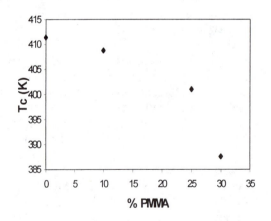

Figure 9 : Variation of fluorinated copolymers crystallisation temperature in PMMA-PVDF-(VDF-TrFE 50-50) ternary blend

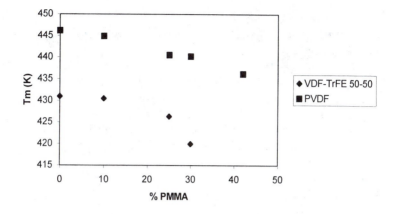

Figure 10 : Variation of fluorinated copolymers melting temperatures in PMMA-PVDF-(VDF-TrFE 50-50) ternary blend

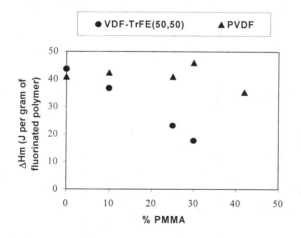

Figure 11 : Variation of fluorinated copolymers melting enthalpies in PMMA-PVDF-(VDF-TrFE 50-50) ternary blend.

128

References

1. Campbell, A.C. *Proceedings OPTO 85*, session V, 117 (1985)
2. Harmon, J.P. 218th ACS Meeting, *Polymer preprints*, **1999**, *40 (2)*, 1256
3. Gillot, C. private communication
4. *Modern Fluoropolymers – High Performance Polymers for Diverse Applications*; Scheirs J., Wiley J. and Sons, Eds; **1997**, *Chapter 26*
5. Faria, L.O.; Moreira, R.L. *Polymer,* **1999**, *40*, 4465
6. Nishi, T.; and Wang, T.T. *Macromolecules,* **1975**, *8*, 809
7. Kwei, T.K.; Patterson, G.D.; Wang, T.T. *Macromolecules,* **1976**, *9*, 780
8. Morra, B.S.; Stein, R.S. *J. Polym. Sci., Polym. Phys. Ed.* **1982**, *20*, 2243
9. Hirata, Y.; Kotaka, T. *Polymer J.* **1981**, *13*, 273
10. Roerdink ,E.; Challa, G. *Polymer* **1978**, *19*, 173
11. Coleman, M.M.; Zarian, J.; D.F.Varnell ,D.F.; Painter, P.C. *J.Polym.Sci., Polym.Lett.Ed.* **1977**, *15*, 745
12 Bertolucci; P.R.H.; Harmon, J.P. "Photonic and Optoelectric Polymers", S.A.Jenekhe and K.J.Wynne, Eds, *ACS Symposium Series,* **1997**, *672*, 79
13 Jouannet, D; T.N.Pham, T.N.; S.Pimbert ,S.; Levesque, *Polymer*, **1997**, *38*, 5137
14 Perrin ,P.; Prud'homme, R.E. *Acta Polymer.* **1993**, *44*, 307
15 Kwei, T.K.; Frisch, H.L.; Radigan, W.; S.Vogel, S. *Macromolecules* **1977**, *10*, 157
16 Del Rio , C.; Acosta, J.L. *Polymer Intern.*, **1993**, *30*, 47
17 Iruin, J.J.; Eguizabal, J.I.; Guzman, G.M. *Eur.Polym.J.*, **1989**, *25*, 1169

Chapter 9

Hard Plastic Claddings: Nearing Two Decades of Performance

Bolesh J. Skutnik[1]

B J Associates, 51 Bunbury Lane, West Hartford, CT 06107
[1]Current address: Fiber Optic Fabrications, Inc., 515 Shaker Road, East
Longmeadow, MA 01028

Hard Plastic Clad Silica (HPCS) optical fibers were invented about two decades ago to improve on the original Plastic Clad Silica (PCS) optical fibers. General properties which materials need to have in order to be good cladding materials for optical fibers are discussed. Details of the invention/innovation process for HPCS are reviewed along with the development of this new type of optical fiber structure. A compilation of the several types now offered in the USA, Japan and Europe is presented. Material suppliers are identified. The advantages and limitations of these fibers for a wide range of medical applications are reviewed. Finally future developments and expanded products are suggested.

Introduction to Polymer Clad Silica Optical Fibers

Before detailing the development and results of Hard Plastic Clad Silica fibers, we will review earlier polymeric cladding materials and in general discuss the requirements on a material to be useful as a cladding.

The key optical properties a polymeric material needs to qualify as a fiber optic cladding are the following two:

(a) the refractive index, n_i, of the polymeric material must be lower than that of silica core material at the operating wavelength and temperatures; and

(b) the optical transmission of the polymeric material should be high over the range of operating wavelengths and temperatures, i.e. low attenuation under operating conditions.

Corollary properties to these are:

(c) the rate of change in refractive index with temperature must not be too extreme compared to that of the silica core material;

(d) lack of significant dimensional transitions over the operating temperature range, even secondary or tertiary ones;

(e) regardless of whether it is physical, mechanical or chemical, good adhesion to the silica core material is desired; and

(f) ability to be applied and 'cured' in line with the drawing of the silica core material into optical fiber.

Many polymeric materials were investigated during the 1970s and early 1980s as possible candidates for fiber optic claddings. Since the refractive index of the polymeric cladding must always be below that of the core material, silica here, there are very few materials that can satisfy this condition. Silicones and fluorinated polymers are the primary candidates, even though many low molecular weight organics have refractive indices below that of silica. The later being 1.4356 at the sodium D-line.

With the exception of thermally cured silicones most of the other materials never routinely appeared in commercial optical fibers. Early researchers believed a soft buffer layer was needed between the optical fiber and a tougher outer jacket. In many ways the soft silicones, with their refractive indices, generally good transparency, and often very low glass transition temperatures, seemed a good answer.

Early on the preference for methyl silicones over phenyl containing silicones was seen. The refractive index of phenyl silicones is higher than that of methyl ones and it rapidly increased to that of pure silica as the temperature approached $0^{0}C$ or below. Secondly, throughout the normal operating range the transmission of methyl silicones was generally better than for the phenyl containing silicones. Plastic clad silica (PCS) optical fibers rapidly came to denote silicone clad silica fibers.

The limitations of the PCS optical fibers soon became apparent as compared to silica/silica optical fibers. PCS optical fibers, especially those with hard outer jackets, increased in attenuation as operating temperature dropped, particularly below $-20\ ^{0}C$. In fact the optical fibers became black, i.e. non-transmitting, at temperatures approaching $-55\ ^{0}C$. The increase in attenuation with decreasing temperature was primarily due to a dropping numerical aperture of the core/clad couple which forced more core modes to become cladding modes that were more easily stripped by the outer jackets.

A somewhat more insidious problem with the silicones is related to their softness. To properly terminate optical fibers, generally a connector must be attached to the fiber's end. For best performance the end usually must be polished to a high degree. These points lead to the need for a solid mating of the connector and the optical fiber so that chatter free polishing may be accomplished. Compression of the silicone in the termination is to be avoided because this would haphazardly increase the refractive index, possibly raising it over that of the silica core. Bonding to silicone is also very difficult because of its low surface energy. Standard termination procedure came to require the removal of the silicone at the termination and the application of a epoxy or other adhesive to bond the optical fiber to the connector. As time went on it became apparent that often it was extremely difficult to get reliable, consistent bonding of the fiber and connector. The culprit was determined to be residual low molecular weight silicone on the optical fiber after the majority of silicone was stripped off. Not only bonding was affected. In line connections appeared to degrade over time, increased attenuation, especially when high power light transmission and/or low pressure conditions existed. The low molecular weight silicone would be drawn into the connection space between the fiber ends; deposited on the ends; and then degraded as high power light became partially absorbed in the coating.

Silicones won out over the fluorinated materials of the time, because they could be applied inline as the fiber was drawn, they could function as buffers, and they did adhere modestly well to the silica surface. Most fluorinated materials were polymers that needed to be extruded onto the forming fibers which meant handling the silica fibers before any protective coatings were applied. This guaranteed a weak fiber with significant surface flaws. Secondly, the adhesion of most of the fluorinated polymers to a silica surface was quite poor. Lastly, because of many applications, which called for high to ultra high purity silicones, the latter were cleaner and thus generally had lower attenuations, higher light transmission, than the fluorinated materials. It was in this situation that the following innovations and inventions were conceived and developed.

Invention/Innovation of Hard Clad Silica Optical Fibers

The concept of a hard plastic clad silica optical fiber arose in 1979. Within the Corporate Research Division of Ensign-Bickford Industries, Inc. [EBI], development work on the standard plastic clad silica [PCS] optical fibers - with soft silicone plastic as the cladding - had progressed to the point where the low temperature optical limitations and the problems with connectorization of the PCS fibers were seen to limit the use of these optical fibers in many applications. Silicone was used because its refractive index, as required for claddings, was below that of the pure silica used in the core of the optical fibers. Soft silicones seemed to be ideal to serve as a buffer coat as well as cladding for silica cores, protecting the latter from mechanical damage. While several silicones have very low glass transition temperatures, where they become hard, even the best of these materials have a minor secondary transition in the -40 to $-50^\circ C$ range that unfortunately greatly increases their refractive index within this temperature range. The attenuation of the fibers thus rises very fast through this temperature range and the fiber becomes unusable near $-50^\circ C$. To achieve good, thermally stable terminations, problems arise from the need to remove the soft cladding from the fiber and the need to remove all the residual material from the silica surface without compromising the surface.

Following conception, I set as the initial design goals for the new optical fibers that they should: (a) have as low an attenuation as possible; (b) have a large numerical aperture, NA, which could be varied; (c) be insensitive to temperatures, particularly to $-50^\circ C$ and below; (d) be as radiation hard as PCS fibers; (e) be easier to reliably terminate than PCS fibers; and (f) be capable of being drawn faster than PCS fibers at that time. Each of these goals could be correlated with one or more cladding material properties. Table I presents the results, obtained with Hard Clad Silica [HCS®] optical fibers, for each of the design/cladding property goals.

Table I. Cladding Property Design Goals/Results

Desired Product Property	Cladding Material Property	HCS Fiber Property
Low Loss	Very Low Absorption 600-1000nm, Tightly Cured Comp.	5-12 dB/km at 820nm
Large NA	Low Refractive Index	NA = 0.38 or 0.45
Temperature Insensitive	Hard, Glassy, Tg> Room T or Very Broad Tg	1 dB/km Added Loss at -50°C
Radiation Hard	Not Interfere with Pure Silica Core Property	Induced Losses Below 10 dB/km at 10krads
Better Connectorization	Cure to Hard, Adherent Glassy Material	Crimp/Cleave Possible
Faster Processing	UV-Curable Composition	2-3 Times Faster than Typical Silicone PCS

Preliminary experiments began in 1980 and by 1981-82 the first HCS fibers were produced and sold. A market study was commissioned and the results indicated a potentially expanding market compared to the standard PCS fiber. The new HCS fibers were described in papers[1,2] presented at the Society of Plastic Engineers ANTEC '83 and at the Optical Society of America CLEO '83 meetings. Besides the properties described in Table I, HCS fibers had an additional structural feature that for all core sizes the cladding thickness was only 7 to 20 μ. This feature as well as a few others, which we will describe later, is important for many medical applications.

Returning to the innovation story, Ensign-Bickford Optics Company was formed in 1984 to produce and market HCS optical fibers. The basic patent for HCS fibers issued in early 1985 and was followed by a second patent in 1987 that described an improved HCS fiber and an HCS-type coated all-silica optical fiber. Having established a beachhead, other workers in the United States and in Japan began work on their own versions of HPCS optical fibers, which culminated in 3M Specialty Optical Fibers obtaining a patent for their TECS® optical fiber and in Dainippon Ink & Chemical Company obtaining a patent for their improvement on the basic HPCS cladding, both in 1989. The latter has given rise to a variety of HPCS type fibers in Japan and Europe. The information on United States patents is given in Table II.

Table II. US Patents for HPCS Materials

Number	Date Issued	Fiber/Producer
4,511,209	April 16,1985[3]	HCS® /Ensign-Bickford Optics [EBO]
4,707,076	Nov. 17, 1987[4]	HCS® +HCS® All-Silica Fibers/ EBO
4,852,969	Aug. 1, 1989[5]	TECS® / 3M Specialty Optical Fibers
4,884,866	Dec. 5, 1989[6]	HPCS generic fibers/ DIC material
5,203,896	April 20, 1993[7]	Optran®/CeramOptec Industries
5,302,316	April 12, 1994[8]	HPCS generic fibers/DIC material
5,690,863	Nov. 25, 1997[9]	?/Optical Polym. Res. material

The formulations for the various HPCS materials are proprietary, but as these patents describe the common thread is to use highly fluorinated esters of acrylic or methacrylic acid with crosslinking multi-functional acrylates and methacrylates, generally photinitiators and commonly adhesion promoters. In some cases catalysts or thermal curing additives have been included in the cladding precursor materials Monomers and oligimers of the fluoroacrylates are used as the base polymer forming component. The adhesion promoters, where present, generally are materials which bond easily to silica surfaces and are compatible or attracted to the basic polymeric backbone.

Properties of HCS® Optical Fibers

HCS optical fiber is of the step-index type. A pure silica rod is drawn to the desired core diameter, then the proprietary hard polymer optical cladding is immediately applied to the pristine surface of the silica core. The polymer is cured with uv radiation and bonded to the silica surface. The cladding is very hard [Shore D 70] and since it is bonded well to the silica surface, it need not be removed when terminated. The cladding is typically about 10μ in thickness which leads to very high core/clad ratios, from 80% to 94%. These three properties - hardness, adhesion to core, high core/clad ratio - are the unique structural properties of HCS and the whole class of HPCS fibers.

A number of mechanical and optical properties differentiated the original HCS fibers from other fibers available at that time. Among the most significant are high tensile strength and high resistance to static fatigue[10]. Dynamic tensile strength for 10m gage lengths is typically above 750 kpsi (>5 GPa) with very high Weibull slopes, $m \geq 60$, i.e. extremely sharp Weibull strength plots. This provides safety margins in critical applications such as medical ones. The static fatigue parameter, n, which is related to the slope of log/log plot of time to failure versus failure stress, is quite high even for fibers immersed in water and lies typically in the 25-30 range. This allows the fiber to safely sustain tight bend radii in applications, as for example in entry to blood vessels through cannula.

The high core/clad ratio of HCS fibers is typically 87% to 94% for the fiber sizes employed in biosensing, endoscopy or laser surgery. Its main advantages are to allow a higher energy density for a given core diameter, provide increased coupling efficiency and to allow easier connectorization. The low attenuation over a reasonably broad wavelength region, down to 4 dB/km @ 800nm, makes the fibers useful for sensitive biosensing applications and for a number of medical laser sources.

HCS fibers can have numerical apertures [NA] of about 0.4 or 0.5 due to the low refractive index of the hard, highly fluorinated character of the cladding. The high NA provides two main benefits, increased coupling efficiency and low bending loss sensitivity. The high coupling efficiency is especially useful for biosensing and illumination applications and the good bending loss sensitivity can be important for all medical, military and industrial applications. The NA does decrease over length as it does for all polymeric clad optical fibers to an equilibrium value, but for the typical 3-5 m lengths used in medical applications the NA value remains constant and high.

The hard, adherent polymeric cladding provides several benefits to HCS fibers. Pistoning in the connector is prevented and the cladding protects unbuffered sections from damage. Since the cladding isn't removed during termination, the reliability of connectors is enhanced. Finally because of these properties it is possible to make reliable crimp & cleave connections, which have the potential of reduced cost and automated assembly.

The HCS optical fibers function well over a wide range of temperatures. Samples of fiber exposed to liquid nitrogen temperatures, -196C [77K], and below were used to carry spectroscopic information from materials held at these temperatures[11]. In the other direction, HCS fibers are stated to be usable up to 125°C, essentially continuously. Of more interest for medical applications is the fact that the fatigue behavior of these fibers, even in highly moist environments, remains predictable and unchanged[12] from ambient water to steam exposure at a temperature of 121°C. The strength at the higher temperatures is reduced somewhat but in a predictable fashion, and it remains high as compared to other fiber types.

The radiation behavior of the standard HCS fibers and special radiation grade fibers are essentially related to the silica materials used in the core[13]. Short and longer term transient behavior have been reported at 865nm[14] and at 1300nm[15] under a number of environmental conditions. These tests were primarily done for military applications, where the dose levels are much lower than for sterilization. All pure silica core fibers with good behavior in this region typically gain little or no additional attenuation as the dose level goes to the 2.5 megarad sterilization dose, assuming that the fiber is illuminated for a period before its final use.

Producers/Properties of Other HPCS Optical Fibers

As mentioned earlier, within the first few years several more producers of HPCS type fibers have come on the scene. Each has their own variation of the cladding as indicated by the patents listed above in Table II. The main players from around the world are HCS® by Spectran Specialty Optics and Toray Fiber Optics; Optran® by CeramOptec Industries and TECS® by 3M Specialty Optics. In Japan the major emphasis has been fiber for the data link and power supply applications, rather than any significant effort in medical applications.

Recently, Fiber Optic Fabrications has begun evaluating the materials available through Optical Polymer Research (OPR). These new materials are available in a variety of viscosities and refractive indices. Preliminary results on clad only optical fibers point to some equivalency with other HPCS clad only optical fibers. Initially the optical properties of the OPR clad fibers, while similar to other HPCS fibers, are not equivalent. Testing and evaluation continues.

Advantages/Limitations for Medical Applications

A number of the properties of HPCS fibers make them very desirable for medical applications, both for laser surgery and for in-Vivo sensing/endoscopy. These are summarized in Table III for the common advantages. High core/clad ratios, low loss, small/large core sizes and broad transmission window are of value for different reasons in each of the application areas. Numerous papers, primarily by EBOC and by 3M at SPIE meetings, have presented and discussed how the different properties are important in various medical applications[17-26].

Table III

Advantages of HPCS Fibers for All Medical Applications

High Inherent Strength	Tighter Bends Permitted in Use
and	and
Good Fatigue Behavior	Packaging & Storage Easier
Hard, Thin, Adherent	Apply Connectors Directly Over Clad
Cladding	Fiber; Improved Reliability
Crimp & Cleave Connectors	Potentially Less Expensive,
on Proximal End	High Volume, Disposable Devices
Cladding Non-Thrombogenic	Diminished Danger in Long Operations
Low Bioload Materials	Easier to Sterilize & to Keep Sterile

As this table shows there are many good reasons for choosing HPCS optical fibers for use in medical devices. However, as with all things, there are also limitations to the currently available products. These are summarized below. In the next section we shall mention developing products, which are addressing some of these limitations.

In laser surgery applications there are two primary limitations. At very high laser powers, the cladding will vaporize, particularly near the fiber ends. This limitation is due to the thermal stability of the proprietary cladding material, to the planarity of the end face, and to minor misalignment of the laser/fiber interface. The ultimate limiting factor is the first one, because to improvements
in the mechanical problems will eventually be limited to the material thermal properties. The use of lasers operating at wavelengths below 400 nm, the near uv region, is hampered by the fact that the hard cladding generally absorbs light strongly in this region of the spectrum. Since most hard claddings are uv cured, they contain uv absorbers, i.e. the photo initiators which start the cure process reactions. Other components of the hard clad compositions may also have major uv absorption characteristics. The absorption in this region increases much more rapidly for the hard cladding than for silica, as is true for most plastic materials.

In the biosensing/endoscopic applications, there are three areas where limitations become evident. A large NA is advantageous for these applications, however for the higher NA HPCS fibers, the hardness, wear resistance and overall robustness typically is reduced from that of the "standard" NA product, making these fibers more susceptible to damage from handling. This is thought due to the need to employ very highly fluorinated components, and the limited availability of such chemicals, particularly as polyfunctional acrylates. Another limitation to increased sensitivity is that for core dimensions smaller than 200 μ the attenuation begins to increase rapidly across the spectrum, because more power travels in the cladding as the core size approaches or drops below 100 μ. The hard plastic clad is more lossy than silica and physics indicates that the signal can extend beyond even the cladding when it is thin and the core size drops below 100 μ. Finally some biosensor systems cannot adhere well to the cladding. This may require the removal of the cladding which is hampered if the cladding adheres too well to the silica core.

Summary

Hard plastic clad silica optical fibers were invented nearly two decades ago in an attempt to achieve the benefits of both all silica and silica/silicone fiber constructions while minimizing each type's problems. To a great extent the original HCS® fibers did indeed meet these intentions. The first aim in 1981 was to provide plastic clad pure silica core fibers with improved thermal and optical behavior for military markets. The thin, hard clad structure was proven to yield high strength fibers with reliable and predictable fatigue properties. This lead to expanding the market to datacom and particularly to medical applications. The mechanical, optical and structural properties have been found to be especially useful in the design of laser fiber delivery systems and in the design of endoscopic and biosensing systems. From one nascent supplier with only two variations, HPCS fibers are now produced in varying 'flavors' by suppliers in the United States, Europe and Japan. New, improved varieties of HPCS fibers are being developed by different fiber manufacturers to expand the capabilities of this interesting class of optical fibers. The invention of the hard cladding with the possibility of formulating changes in many of the optical and thermal properties and thus producing new fibers has vitalized the general plastic clad silica fiber market.

References

1. B.J. Skutnik, *ANTEC '83* Proc.**1983**, 436 (Society of Plastics Engineers).
2. B.J. Skutnik and R.E. Hille, *CLEO '83* Proc., **1983,** (Optical Society of America).
3. B.J. Skutnik, US Patent # 4,511,209 (1985).
4. B.J. Skutnik and H.L. Brielmann, Jr., US Patent # 4,707,076 (1987).
5. S.A. Babirad, F. Bacon, S.M. Heilmann, L.R. Krepski, A.S. Kuzma, and J.K. Rasmussen, US Patent # 4,852,969 (1989).
6. Y. Hashimoto, M. Kamei and T. Umaba, US Patent # 4,884,866 (1989).
7. W. Neuberger, US Patent # 5,203,896 (1993).
8. Y. Hashimoto, J. Shirakami and M. Kamei, US Patent # 5,302,316 (1994).
9. P.D. Schuman, US Patent # 5,690,863 (1997).

10. W.B. Beck, M.H. Hodge, B.J. Skutnik and D.K. Nath, *EFOC/LAN* Proc., **1985** 145 (Information Gatekeepers Inc.).
11. Schwab, S.D. and McCreery, R.L., *Anal. Chem.* **1984,** *56,* 2199.
12. Skutnik, B.J., Hodge, M.H. and Clarkin, J.P., SPIE **1988,** *842,* 162.
13. Skutnik, B.J. and Hille, R.E., SPIE **1984,** *506,* 184.
14. Skutnik, B.J., Greenwell, R.A. and Scott, D.M., SPIE **1988,** *992,* 24.
15. Evans, B.D. and Skutnik, B.J., SPIE **1989,** *1174,* 68.
16. McCann, B.P., SPIE **1991,** *1420,* 116.
17. Skutnik, B.J., Hodge, M.H. and Beck, W.B., SPIE **1987,** *787,* 8.
18. Skutnik, B.J., Hodge, M.H. and Clarkin, J.P., SPIE **1988,** *906,* 21.
19. Skutnik, B.J., Brucker, C.T. and Clarkin, J.P., ibid. 244.
20. Skutnik, B.J., Clarkin, J.P. and Hille, R.E., SPIE **1989,** *1067,* 22.
21. Skutnik, B.J., Hodge, M.H. and Clarkin, J.P., ibid., 211.
22. Skutnik, B.J., SPIE **1990,** *1201,* 222.
23. McCann, B.P., *Photonics Spectra* **1990,** *24* (5), 127.
24. Krohn, D.A., Maklad, M.S. and Bacon, F., SPIE **1991,** *1420,* 126.
25. McCann, B.P. and Magera, J.J., SPIE **1992,** *1649,* 2.
26. Skutnik, B., Neuberger, W., Castro, J., Lashinin, V.P., Blinov, L.M., Konov, V.I. and Artjushenko, V.G., ibid., 55.

Chapter 10

UV Curable Acrylated Oligomers: Model Characterization Studies

A. J. Tortorello

DSM Desotech, Inc., 1122 St. Charles Street, Elgin, IL 60120

Unique acrylate functional oligomers have been developed with the concept of versatile property selection in mind. When formulated into typical UV curable coatings for optical fiber, a range of liquid and film properties become attainable depending upon the oligomer being used. Cured films covering the spectrum from soft and flexible inner primary types to hard and brittle outer primary types have been made.

This study examined the characterization of these oligomers. In particular, reactions of model compounds were used to assist in structural confirmation especially with respect to IR functional group absorptions and NMR chemical shifts. The study of these model compounds has provided reliable evidence that the required chemical functionalities are part of the desired oligomer.

Introduction. Acrylic copolymers have been used in the coatings industry for many years.[1] This is particularly true for architectural and automotive finishes where exterior durability is at a premium. Their use has been broadly documented in a variety of volatile media including organic solution coatings (lacquers), water based dispersion coatings, and water based latex coatings. The attraction of acrylic copolymers is many-fold. Monomers used to synthesize the polymers are based on readily available feedstocks; reactions are well known and can be easily run in conventional production equipment; the copolymers have demonstrated excellent exterior durability in resisting the effects of exposure to south Florida sunlight. Perhaps the greatest attraction, however, is in the variety of properties these copolymers can provide. This versatility ranges from reactive functional groups for subsequent curing chemistry to glass transition temperature selection to molecular weight design and rheological control.

Acrylics are commonly used in the radiation cure coatings industry as well. Acrylate-functional monomers and oligomers are copolymerized in place on a substrate through the activation of a UV sensitive free-radical initiator.[2] Typical acrylate-functional oligomers are derivatives of other functionalized low molecular weight prepolymers such as the diglycidyl ether of bisphenol-A, ethoxylated bisphenol-A, the hydroxy-terminated polyethers of propylene oxide and tetrahydrofuran, hydroxy-functional polysiloxanes, and hydroxy-terminated polyesters. Many of these are commercially available from a variety of suppliers.

Providing a higher molecular weight acrylic copolymer with a pendant or terminal acrylate functionality has not been widely practiced because of the unique synthetic challenge it offers. The acrylate functionality in one part of the polymer molecule must be preserved during reaction of the identical functional group used to synthesize the polymer. Clearly, the acrylate functional groups cannot be part of the same monomer since multi-functional acrylates are well known crosslinking agents. Alternatively, a suitably functionalized polymer could be synthesized from acrylate monomers followed by conversion of the pendant functional groups through subsequent reaction with a suitably functionalized acrylate monomer. This sequence could be facilitated by polymerization in solvent so as to prevent gellation from the Trommsdorf effect. However, the solvent would have to be substantially removed in a subsequent step in order to have a desirable product for use in a UV curable application.

Recently, we have developed a technique to provide compositions which are solvent-free and contain acrylate functionality pendant to an acrylic copolymer backbone.[3] This report summarizes the characterization of the copolymer compositions and the model reaction studies used to assist in structural confirmation.

Experimental. Acrylic copolymer solutions (1) were synthesized by free-radical polymerization of acrylic monomers at elevated temperature. Typically, reactive diluent polyol was heated to 80°C in a flask equipped with reflux condenser, agitator, nitrogen inlet line, and monomer feed line. A solution of mixed monomers and thermal initiator (t-butylperoxy 2-ethylhexanoate) were continuously metered to the flask over three hours. After addition, the solution was further heated to 100°C for an additional three hours and finally cooled before pouring out.

Urethane acrylates (3) were synthesized by initially forming the isocyanate functional acrylate compound (2) *in-situ*, reacting one mole of a suitable diisocyanate (isophorone diisocyanate) with one mole of 2-hydroxyethyl acrylate. The previously synthesized acrylic copolymer solution (1) is then added to the flask contents and the mixture heated to 80°C in the presence of acrylate reactive diluent (isobornyl acrylate) to control viscosity. The temperature is maintained until the isocyanate is essentially all consumed as determined by titration.

UV curable coatings were prepared by mixing urethane acrylate oligomer (3), additional diluent monomer (isobornyl acrylate, isodecyl acrylate, or 2-(2-ethoxyethoxy) ethyl acrylate), and photoinitiator (2-hydroxy-2-methyl-1-phenyl-propan-1-one) at 50°C until homogeneous. The coatings were then cast onto glass plates using a 75 micron controlled thickness applicator. The liquid coatings were cured by exposure to a medium pressure mercury lamp at a doseage of 1 joule per square centimeter. The cured films were carefully removed by undercutting suitably sized strips with a razor.

Results and Discussion. Synthesis of acrylic copolymers in solution phase is necessary to prevent gellation. At the same time, it is also undesirable to have UV curable compositions containing substantial amounts of volatile components not only for the environmental impact but also for the negative effects excessive shrinkage produces in an applied film. Our approach to resolve this apparent conflict was to invoke the concept of reactive diluency. The polymerization could be run in a diluent which is nonvolatile yet could partake in a subsequent reaction to introduce acrylate functionality onto the pendant functional groups. The hydroxyl-isocyanate reaction is conveniently suited to this synthetic methodology. Therefore, hydroxy-functional acrylic copolymers are synthesized in the presence of hydroxy-functional prepolymer diluents. The resulting solution is converted to acrylate functionality by subsequent reaction with an isocyanate functional acrylate compound. This sequence is outlined simplistically in Scheme 1 wherein compound 1 refers to the hydroxy-functional acrylic copolymer solution in reactive diluent, compound 2 represents the isocyanate functional acrylate compound and compound 3 is the resultant urethane acrylate.

SCHEME1: ACRYLATED ACRYLIC SYNTHESIS

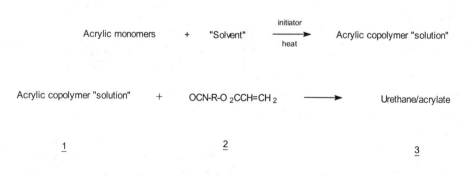

A. Acrylic Copolymer Solution (1) Properties. Much of the versatility attributed to acrylic copolymers arises from the control of glass transition temperature for film property variation. Control of the glass transition temperature is well known and frequently practiced in the form of the Fox equation[4] described as follows:

$$(Tg)^{-1} = \Sigma w_i (Tg_i)^{-1} \quad \text{(Eqn. 1)}$$

where Tg is the copolymer glass transition temperature (in degrees Kelvin), w_i is the weight fraction of monomer and Tg_i is the glass transition temperature (Kelvin) of the homopolymer derived from the monomer.

Effect of Acrylic Tg on Solution Viscosity. Apart from the obvious effect of copolymer glass transition temperature on final cured film Tg, the copolymer Tg is also known to affect solution viscosity. This relationship was originally derived for polymer melt viscosity by Williams, Landel, and Ferry.[5a] In its simplified form, the WLF equation is described as follows:

$$\log (\eta_T/\eta_{Tg}) = -[A(T\text{-}Tg)/(B+T\text{-}Tg)] \quad \text{(Eqn. 2)}$$

where η_T and η_{Tg} are the viscosity (poise) of the polymer at temperature T and Tg, respectively, and A and B are predefined constants. Hill[5b] demonstrated the predictive capability of the simplified WLF equation for high solids coating viscosities where Tg represents the glass transition temperature of the high solids solution. The effect of copolymer glass transition temperature on solution viscosity is demonstrated in Table 1 for a series of methyl methacrylate (MMA), 2-ethylhexyl acrylate (EHA), 2-hydroxyethyl acrylate (HEA) copolymers in 400 molecular weight polypropylene glycol diluent. Viscosity is seen to decrease with glass transition temperature.

Table 1: Effect of Acrylic Copolymer Tg on Solution Viscosity

	A	B	C
Tg (°C, cal'd)	0.3	-19.0	-35.4
Diluent polyol mol. wt.	400	400	400
Acrylic copolymer conc. (wt %)	50.54	50.81	50.81
Hydroxyl value (mg KOH/g)	151.99	139.30	151.18
Color (Gardner)	1	1	1
Viscosity (Pa•s, 25°C)	>1 MM	332.0	57.3

Effect of Copolymer Hydroxyl Content on Viscosity. Since the acrylic copolymers are hydroxy-functional and the prepolymer diluent is also hydroxy-functional, we sought to examine the effect of acrylic hydroxyl content on solution viscosity. These results are summarized in figure 1, which depicts the effect of hydroxyl level in the acrylic polymer on solution viscosity for a series of polypropylene glycol diluents differing in molecular weight only. Generally, one sees a trend to lower viscosity with decreasing hydroxyl content. This can most likely be explained by the effect of intramolecular association probably caused by the polarity of the hydroxyl groups on the polymer backbone.

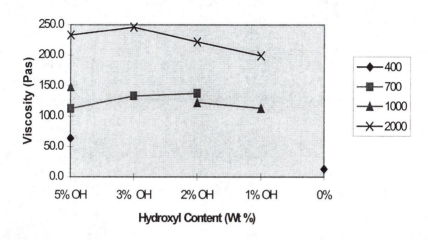

Figure 1: Hydroxyl Content Effect

Acrylic Copolymer Grafting onto Diluent Polyol. The selection of hydroxy prepolymers as diluents for use in the acrylic copolymerization provides another level of versatility in property design but also must be considered judiciously. Table 2 represents a summary of gel permeation chromatography results for a series of identical acrylic copolymers synthesized at the same weight concentration in different molecular weight polypropylene glycol diluents. The reported values refer to only the acrylic portion of the chromatogram, which was well distinguished from the diluent portion. The table indicates a gradual increase in acrylic copolymer molecular weight as the diluent molecular weight increases and suggests that some grafting of the acrylic copolymer onto the polyol may be occurring. This can be explained by recalling the fact that hydrogen atoms on carbon atoms alpha to an oxygen are free-radically labile and thus can be abstracted by active free-radicals. Once abstracted the carbon atom can act as an initiation site for monomer grafting.

Table 2: Effect of Diluent Molecular Weight on Acrylic Copolymer

Diluent Molecular Weight	Mw	Mn	Mw/Mn	Mz
400	87381	18060	4.84	202940
700	90075	21177	4.25	205499
1000	104344	25484	4.09	236939
2000	115992	32404	3.58	261939

B. Urethane/Acrylate (3) Structure Confirmation By Model Studies. During the subsequent synthesis of the urethane acrylate (3) from the acrylic polyol (1), we sought to determine that the acrylate compound (2) was indeed reacting with the acrylic copolymer. We had indirect evidence from end group titration that the isocyanate compound was reacting to greater than 95% conversion, but sought direct evidence that it was actually adding to the acrylic backbone. To assist in this determination we designed a model compound study subjecting the starting materials and products to spectroscopic analysis. Equation 3 depicts the details of this study.

$$P-CH_2CH_2OH \xrightarrow{\text{RCH}_2\text{NCO}} P-CH_2CH_2O_2CNHCH_2R \qquad \text{(Eqn. 3)}$$

$$\text{(a)} \quad \text{(b)} \qquad\qquad\qquad\qquad \text{(a)} \quad \text{(b)} \quad \text{(c)} \quad \text{(d)}$$

Model acrylic copolymers were synthesized in n-butyl acetate solution followed by reaction at 80°C with an alkyl isocyanate until conversion exceeded 95% by titration. The product was then cast as a film on glass and the solvent was evaporated in a vacuum oven. Samples were collected for infrared, carbon-13 and proton NMR spectra. Equation 3 also indicates the designations of nuclei for NMR chemical shift assignments.

Fourier Transform Infrared Structure Confirmation. Figures 2 and 3 are representative and depict Fourier Transform infrared spectra of one model acrylic copolymer and its urethane derivative, respectively. Among the features of note in figure 3 is the absence of an isocyanate absorption at approximately 2270 wavenumbers and the presence of an absorption at roughly 1530 wavenumbers typically assigned a urethane C-N-H bend.

Structure Confirmation by NMR. Table 3 summarizes the resonance assignments of the carbon and proton nuclei for the acrylic copolymers and their urethane derivatives. These assignments were guided by reference to those of a hydroxyl-functional acrylate monomer as a model. We used 2-hydroxyethyl acrylate (HEA) whose assignments are also included in the table. The most distinguishing characteristics of the NMR spectra are the development of a new carbon resonance at 155.8 ppm attributed to the urethane carbonyl (c), the downfield shift of the carbon alpha to the ester oxygen (b) from 60.4 to 62.7 ppm, the development of a new carbon resonance at 61.8 ppm assigned to the carbon bonded to the urethane nitrogen atom (d), the downfield shift of the protons on the carbon bonded to oxygen (b) from 3.80 to 4.25 ppm, and the development of a new proton resonance at 3.15 ppm assigned to the protons on the carbon bonded to the urethane nitrogen atom (d). The proton bonded to the nitrogen atom is indistinguishable as it is likely buried among the resonances around 2.0 to 2.5 ppm.

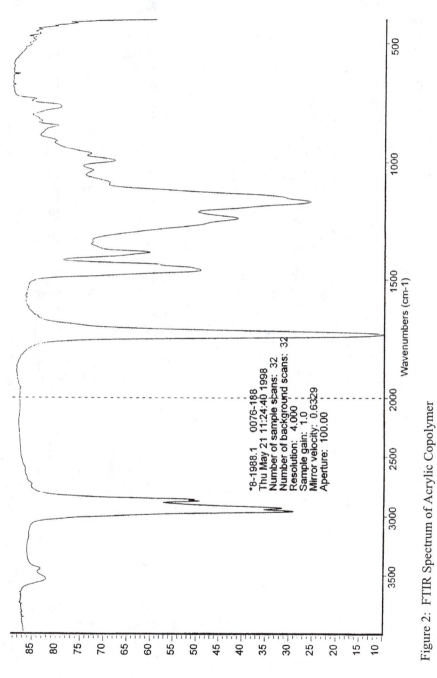

Figure 2: FTIR Spectrum of Acrylic Copolymer

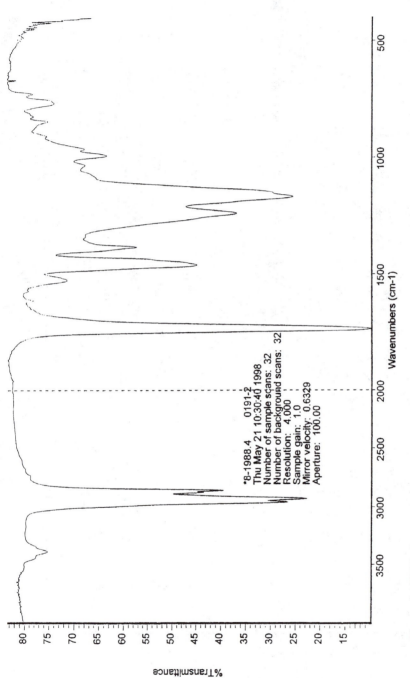

Figure 3: FTIR Spectrum of Urethane of Acrylic Copolymer

Table 3: Summary of NMR Chemical Shift Assignments

Nucleus	HEA		Acrylic		Urethane	
Designation	^{13}C	^{1}H	^{13}C	^{1}H	^{13}C	^{1}H
a	65.1	4.26t	66.8	3.95	66.8	3.90
b	59.3	3.84t	60.4	3.80	62.7	4.25
c					155.8	N/D
d					61.8	3.15

Figures 4 and 5 are the carbon magnetic resonance spectra of the acrylic copolymer and its urethane derivative, respectively. Figures 6 and 7 represent the proton magnetic resonance of the same two materials.

C. UV Cured Film Properties: The urethane derivatives of a number of acrylic copolymers were used in UV curable coating formulations to examine the range of properties achievable from such compositions. Formulations were designed to evaluate the effect of different oligomers and different reactive diluent monomers on cured film properties. To simplify the comparisons, the oligomer concentration, the diluent concentration, and the photoinitiator type and concentration were all maintained at constant levels. A representative composition would be oligomer (40 weight pct.), reactive diluent monomer (60 weight pct.), and photoinitiator (3 pph).

Figure 8 depicts the combined effects of diluent polyol molecular weight from the acrylic copolymer synthesis and the glass transition temperature of the reactive diluent monomer in the UV curable formulation on cured film modulus. The results are not especially surprising and predict that modulus will increase with reactive diluent Tg and decrease with diluent polyol molecular weight.

Figure 9 represents the combined effects of the same two variables on cured film elongation. Once again the results are not that surprising and predict that elongation will decrease with increasing reactive diluent Tg and will increase with diluent polyol molecular weight especially when high Tg diluent monomers are used.

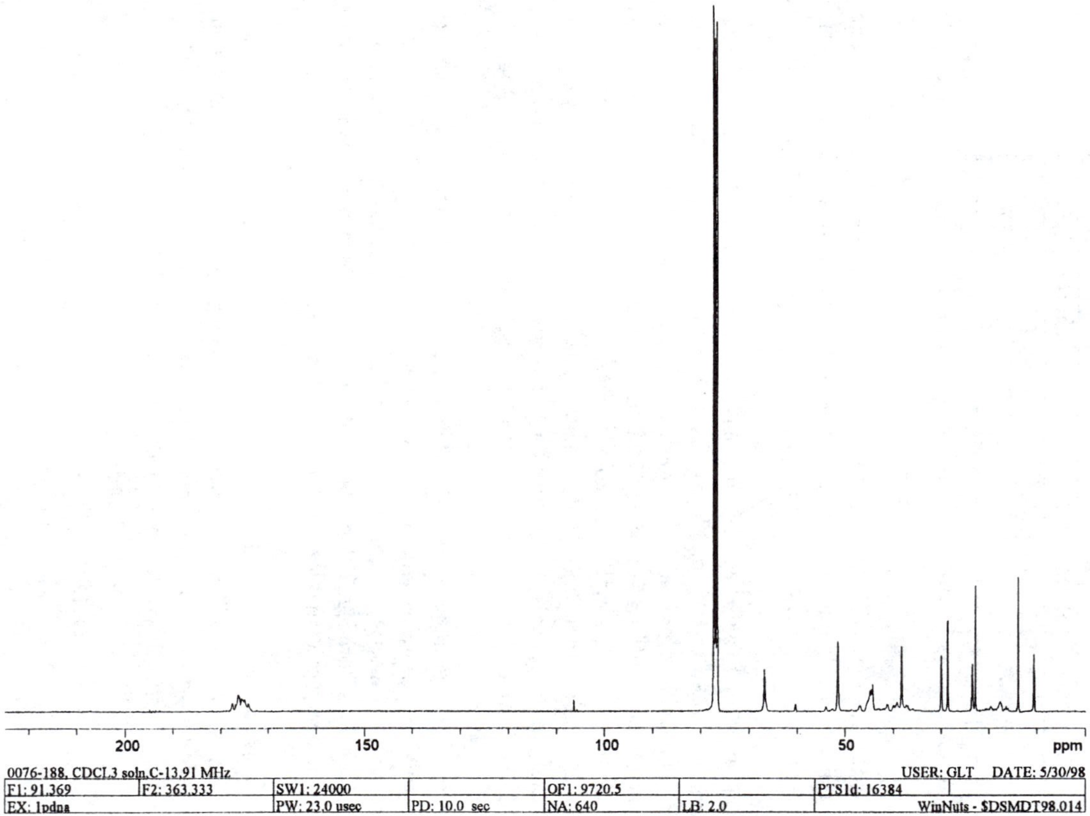

Figure 4: CMR Spectrum of Acrylic Copolymer

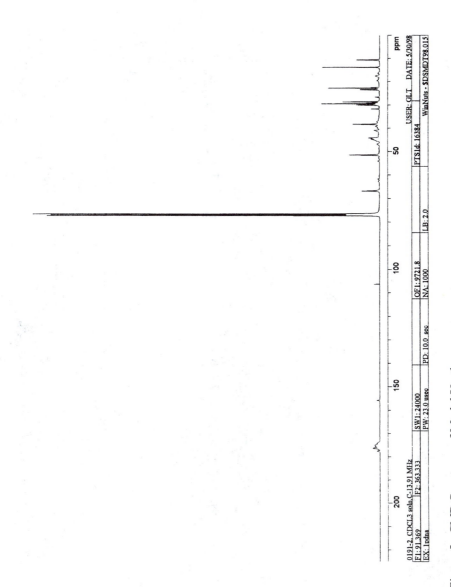

Figure 5: CMR Spectrum of Model Urethane

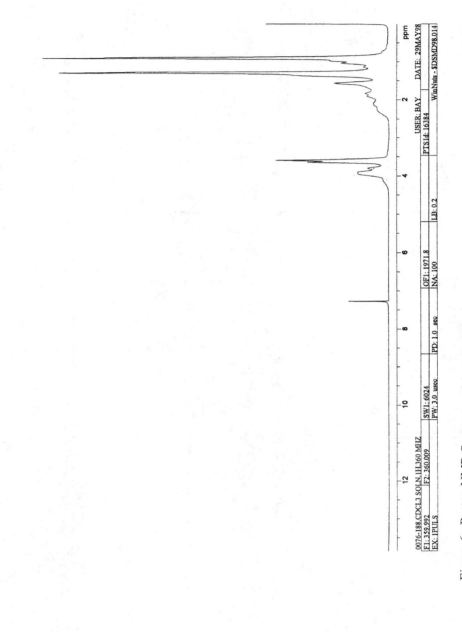

Figure 6: Proton NMR Spectrum of Acrylic Copolymer

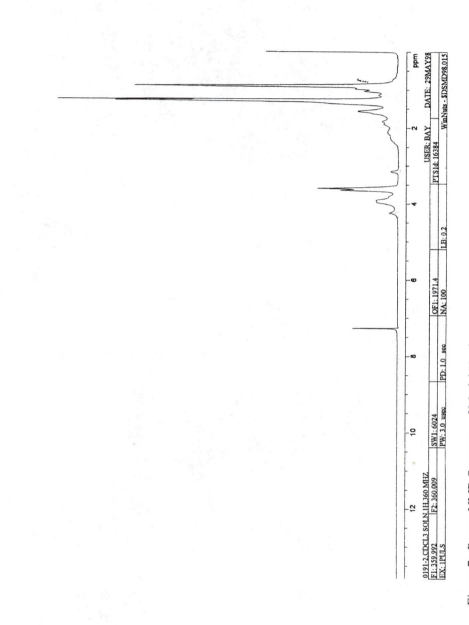

Figure 7: Proton NMR Spectrum of Model Urethane

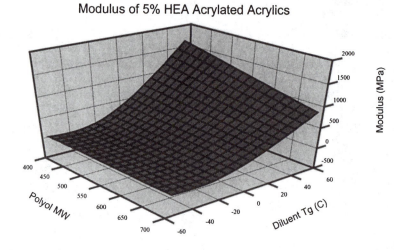

Figure 8: Combined Effects on Modulus of 5% HEA Acrylics

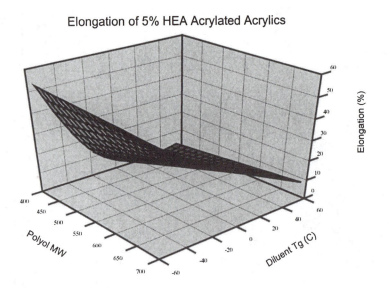

Figure 9: Combined Effects on Elongation of 5% HEA Acrylics

Conclusions. The synthesis of urethane oligomers bearing acrylate functional groups from acrylic copolymer starting materials has been demonstrated. These starting materials offer considerable versatility in design of properties because of their range in glass transition temperature, diluent selection, and functionality level. Effects of variation in these design characteristics on material properties have been determined. Chemical modification of these starting materials has been demonstrated using spectroscopic structural confirmation studies on model compounds. The oligomers were used in prototype UV curable coating formulations with the intent of examining the effect of compositional variation on cured film properties.

References.

1. a) Brown, W.H.; Miranda, T.J. *Official Digest*, **1964**, *8*, 92-134; b) Brendley, W.H. *Paint & Varnish Prodn.*, **1973**, *July*, 19-27.
2. a) Pappas, S.P. ed. "UV Curing: Science & Technology;" vol. 2, Technology Marketing Corp., Norwalk, Conn., **1985**, ch. 4&5; b) Costanza, J.R.; Silveri, A.P.; Vona, J.A. "Radiation Cured Coatings," in Federation Series on Coatings Technology, Brezinski, D.R. and Miranda, T.J., eds. Federation of Societies for Coatings Technology, Philadelphia, PA, **1986**, 7-24.
3. Tortorello, A.J.; Murphy, E.J. US 5847021, December 8, 1998; (DSM N.V.)
4. Fox, T.G. *Bull. Am. Phys. Soc.*, **1956**, *1*, 123.
5. a) Williams, M.L.; Landel, R.F.; Ferry, J.D.; *J. Amer. Chem. Soc.*, **1955**, 77, 3701; b) Hill, L.W.; Kozlowski, K.; Sholes, R.L.; *J. Coatings Technol.*, **1982**, *54*(692), 67-75.

Chapter 11

Micro Porous Silica: The All New Silica Optical Fibers

Bolesh J. Skutnik

Fiber Optic Fabrications, Inc., 515 Shaker Road, East Longmeadow, MA 01028

A new type of optical fiber has been developed. It is made with all pure silica in both the core and cladding. This is possible because the cladding is a micro porous silica produced by a modified sol-gel technology. The formation and characteristics of this new optical fiber type are described. In particular the optical and mechanical properties are illustrated and described. The strength and fatigue of these optical fibers are very good, even without additional protective jackets. Unjacketed fibers have mean Weibull strengths in bending of 6.5 to 7.6 GPa with Weibull slopes in the 40 to 60 range. Fatigue results for fibers tested in ambient air, ambient water and boiling water are presented. The dynamic and static fatigue parameters are around 20. The micro porous structure of the sol-gel cladding provides sites for attaching different moities which could activate biochemical reactions or be useful as sensing sites. Based on preliminary experiments, some possibilities are presented. In general this new structure can provide opportunities for as yet unidentified applications where chemicals and or light must be brought in close contact with body tissue to effect beneficial reactions there.

In 1993 a new approach to "sol-gel" material processing was discovered[1]. Micro porous silica could be produced on a pure silica core, such that a truly functional optical fiber was fabricated[2]. These experiments were an outgrowth of work begun in the 1980s[3-5]. Objectives of this research are to complete feasibility studies on the development of lower cost, ultra-long (>50 km) single mode optical fibers and to develop radiation resistant multi mode optical fibers for extra-terrestrial applications, such as space satellites. A secondary objective is to establish the advantages of a micro porous clad pure silica core optical fiber as a specialty optical fiber for new applications.

The micro porous clad fibers behave like hard plastic clad silica (HPCS) optical fibers[6] in that they do not require a buffer or jacket over the cladding. As with HPCS fibers, however, a jacket is beneficial for most practical environments[7,8].

The optical and the mechanical properties of the micro porous silica clad optical fibers, including the strength and fatigue properties are presented. Potential new applications within medical and biochemical fields are suggested.

Experimental

The material used as a precursor is a hydroxy, ethoxy terminated ladder oligomer with significant Si-C linkages as methyl groups attached to silicon[9]. Details of the materials and application scheme appear below.

Chemical Details

The material used as a precursor is a hydroxy, ethoxy terminated ladder oligomer with significant Si-C linkages as methyl groups attached to silicon as shown in Figure 1. This spin glass resin, GR650, is manufactured by Techneglas of Toledo, Ohio. It was applied from a solution in ethyl acetate, generally at a 50/50 solute/solvent weight percent ratio. A polymeric jacket, when applied, consisted of a single coat of a polyimide, typically a HD Microsystems product, applied in line but on a coating wall after the optical fiber came off the draw tower. A material similar

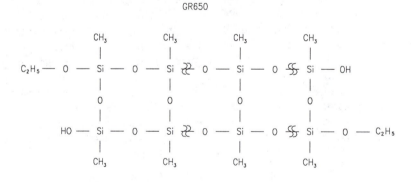

Figure 1: Idealized chemical structure of the glass resin used as the sol-gel precursor for the cladding. This oligomer is end capped with hydroxy and ethoxy groups and has one methyl group per silicon atom.

to GR650, hydrogen silsesquioxane, whose molecular structure was similar but which had to be applied as a dispersion, was used early in the work[2]. It had several problems, in general including having to apply dispersions with less than 10% solids, and gave more scattered results.

Processing Details.

The micro porous silica clad optical fibers were drawn on conventional fiber draw towers[10] using thermal ovens to cure/convert the precursor materials to the micro porous silica cladding. The majority of the fibers were drawn on a standard 8 meter tall fiber draw tower with up to four ovens available in line. Typically temperatures in the range of 450 to 700 ^{0}C were employed. Some of the fibers were drawn on a 12 meter tall fiber draw tower which was equipped with six ovens in line.

The cure and conversion of the silica precursor to micro porous silica is done in line. The general process is that solvent is driven off as the hydroxy and ethoxy groups react to create the cured polymer. The water and ethanol residues flash off. Continued exposure to greater than 400 ^{0}C causes oxidation of many of the remaining Si-C bonds (methyl groups). The conversion is not complete under current conditions. Draw speeds of 7 to 70 meters/min have been employed to date. Typical residence times in a given oven is thus 0.5 to 5 seconds with a total time of exposure to elevated temperatures of approximately 5 to 30 seconds. In line processing refers to processing before the optical fiber is diverted from vertical travel, i.e. before the fiber comes in contact with the pinch wheel at the bottom of the draw tower.

For the results described below, a double application was used with a final sol-gel cladding thickness of ~12 µm. The size of the solid silica core varied among the runs from diameters of about 100 µm to about 200 µm. In a few cases, a preform made from pure silica core and a fluorosilica cladding was used as the starting point. Clad/core ratios varied from 1.1 - 1.4. The core silica material in all cases was a low OH undoped silica, with the OH level typically less than 5 ppm. When applied, the jacket material was a single coat of a polyimide, cured by a series of ovens ranging in temperature from 300 to 500 ^{0}C. Polyimide thickness was ~7 µm.

The actual dimensions of specific fibers are given with the figures displaying results. The core silica material in all cases is a low-OH undoped silica, typically less than 5 ppm of OH. When applied, the jacket material was a single coat of a polyimide, cured by a series of ovens. Polyimide thickness was ~7 µm.

Optical Testing Details

The primary tool in making the optical measurements is an Optical Time Domain Reflectometer (OTDR) having a source emitting at 850 nm. The signal is transmitted to the fiber under test usually through a pigtailed optical fiber, having similar core dimensions and numerical aperture (NA) to the test fiber, which is approximately 0.25 for the sol-gel clad fibers in the current study. A portion of the signal is reflected back as it travels down the test fiber. The OTDR measures the loss in this reflected signal in comparison to the launched signal.

Strength Testing Details

Two point bend strength testing has been described in the literature[11-13]. The two point bend tests were performed on a 2-point bend tester with grooved stainless steel faceplates holding up to 10 fibers at a time. Strain rates and other specifics are given with the figures.

Note that the gage length for this type of test is small, on the order of several millimeters. Samples, however, were normally not taken in a row. Rather, 10m to 100m sections of fiber were produced between sets of ten [10] samples. Additionally many fibers tested were produced under essentially equivalent conditions so that fiber strength measured under ambient test conditions has used total lengths of fiber tested of a meter or more. This sample total is still small but spans a large amount of fiber produced (>20 km) from which the smaller sections of samples have been taken for actual testing.

Fatigue Testing

Static fatigue measurements were performed by the mandrel wrap method[14-16]. The length of fiber under test was about 500 mm. The fibers were wound in ambient conditions and tested in ambient water or boiling water. Stresses lay between 1.4 and 4.6 GPa [200 to 667 kpsi]. A minimum of five samples were used for each mandrel size.

Dynamic fatigue measurements are made in conjunction with the strength testing. By varying the speed of the strength testing, the effects of dynamic fatigue are observed. Typically, if the strength data has a plot without breaks, the mean strength is tracked versus the rate of stress or versus the rate of strain. If there are breaks in the strength plot at a given test rate, then a specific ordered sample strength, one with a specific failure probability, can be tracked versus the change in test rate.

Results

Two-Point Bend Strength Tests.

Figure 2 presents data from several fiber draws where over 220 samples were tested out of >20 km of fiber produced under equivalent conditions over a twelve month period. Except for the very strongest values the curve is rather steep. There appears to be no real breaks in the slope, indicating a single flaw distribution.

Figures 3, and 4 present the change in Weibull strength as the constant strain rate is changed for tests run in ambient air, and in ambient water, respectively.

In each figure the unjacketed fiber results are denoted by Xs, while the polyimide jacketed fiber results are denoted by open circles, Os. As can be seen the jacketed fiber results are generally steeper in slope than the unjacketed fiber results. Curves for the jacketed and unjacketed fibers are very close to one another in ambient water testing (Figure 4).

Static Fatigue Tests.

The results for ambient water immersion include moderately long term failures with all the initial samples of the unjacketed fiber having been broken. In Figure 5 the time to failure for all samples are presented. The data points (diamonds) of an unjacketed fiber clearly show a break in the curve. The short term data in Figure 5

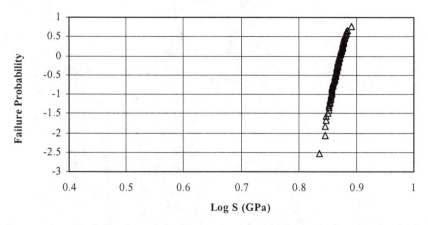

Figure 2: Weibull plot of the dynamic strength for unjacketed sol-gel clad optical fibers, having a pure silica core with diameter of 165 µm and a clad diameter of 195 µm; constant strain rate of 10%/min; 2-point bend test carried out in ambient air, ~23⁰C 50% RH

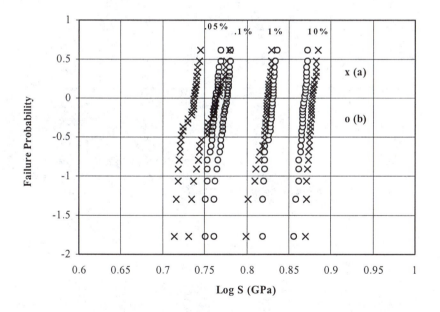

Figure 3: Weibull plots of dynamic strength measurements taken at strain rates varying from 10%/min down to 0.05%/min (a) unjacketed sol-gel clad optical fibers nominally 165/190 µm core/clad diameters, "x" points; (b) polyimide jacketed sol-gel clad optical fibers nominally 204/232/244 µm core/clad/polyimide diameters, "o" points; 2-point bend test; ambient air = 23 ⁰C, 35-60% RH.

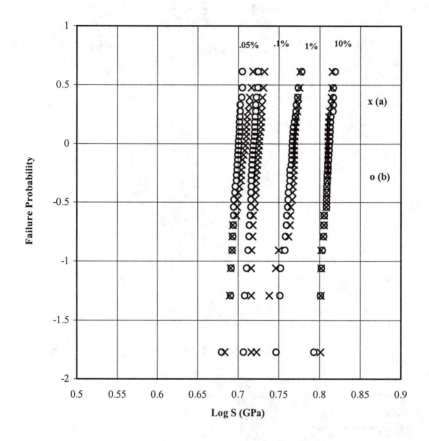

Figure 4: Weibull plots of dynamic strength measurements taken at given strain rates, "x" points for unjacketed fiber of Figure 3(a), "o" points for jacketed fiber of Figure 3(b); 2-point bend test; testing in ambient water, 23 ^{0}C.

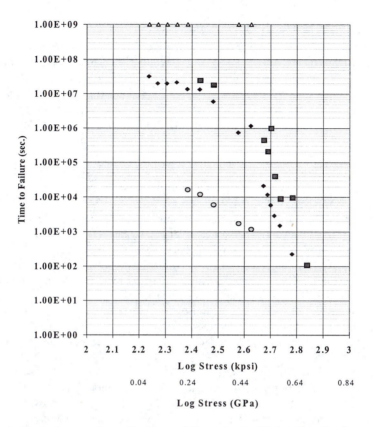

Figure 5: Static fatigue plots of average failure times of 0.5 m length of sample fiber wrapped around mandrels giving the bending tensile stresses plotted, "diamond" points for unjacketed fiber whose dimensions are 164/184 μm core/[sol-gel clad],N_S (hi) = 18, N_S (knee) = 3; "box" points for jacketed fiber whose dimensions are 162/192/208 μm core/[sol-gel clad]/polyimide; testing in ambient water, 23 °C; points at top of graph are fiber samples which have not yet failed; N_S (PI, water) = 25, N_S (PI, boiled) = 6.

indicate a static fatigue parameter, N_S, of about 18 for the unjacketed optical fiber. Above the break at longer times, the static fatigue parameter has dropped to a value of about 3. Further in this figure the upper squares represent the time to failure for the polyimide jacketed optical fiber. Triangular points at the top of the figure represents samples which had not broken when this graph was prepared about 690 days, >16,600 hours [~6 x 10^7 sec], into the testing program. The static fatigue parameter is 21. Also note that the highest stress tested is over 4.6 GPa [660 kpsi].

The lowest set of points in Figure 5, the gray circles, represent the average time to failure for polyimide jacketed optical fibers immersed in boiling water. The plot of these points appears to be linear, though some curvature seems present. The static fatigue parameter, N_S (100 °C), has a value of approximately 6.

Modification/use of micro porous clad structure.

Studies are underway to explore the doping or modification of the micro porous structure of the cladding to provide either active sites for chemical or biological activity or for sensing applications. Preliminary results indicate attachment of Rhodamine complexes which were activated by light energy transmitted down the fiber to the sites where the complexes were 'incorporated'. Also gratings have been formed within the micro porous structure through the use of photo activated precursors.

Discussion

Two-point bend strengths vs. strain rates.

The results in Figure 2 demonstrate the consistency of the results obtained from a large number of fiber draw runs, i.e. a large number of silica rods and sol-gel precursor preparations. The samples tested for this figure represent a random sampling of over 20 kilometers of modified sol-gel clad optical fiber. A main feature of Figure 2 is the steepness of the slope, especially in the upper region, which tends to imply that the flaw distribution may actually be unimodal[17].

The results of Figures 3 and 4 demonstrate the relatively equivalent behavior of the unjacketed sol-gel clad fibers and the polyimide jacketed fibers. At the two slowest strain rates the data for the unjacketed fiber clearly shows a discontinuity in the slope.

Dynamic fatigue results

The dynamic fatigue of unjacketed and polyimide jacketed fibers were determined from the results given above for the two conditions presented. Within experimental error it can be stated that the dynamic fatigue parameter is $N_D = 22 \pm 2$ regardless of whether the sol-gel clad optical fiber is jacketed with polyimide or not.

Static fatigue behavior

By their nature static fatigue measurements usually yield much longer times to failure and thus samples experience much longer exposure to the test environment than in dynamic fatigue testing. Small differences in the fatigue behavior of similar samples become more pronounced in static fatigue testing. Furthermore in studies, like the present one, where dynamic measurements are made by the two point method, the gage length of the static fatigue samples are generally longer than for the dynamic fatigue samples. The two features greatly enhance the observation of differences in sample behavior by static fatigue testing over dynamic fatigue testing. The appearance of a 'knee' in the data plotted in Figure 5 for the unjacketed fiber samples is an example of this phenomenon. This 'knee' is a well recognized feature of the standard telecom optical fibers and many other fibers as well. The research groups at Bell Labs and at Rutgers University[11,19,20,21] and others[22] have reported on the occurrence and significance of this feature. The 'knee' feature was found present in bare fiber work and in most acrylate jacketed optical fibers.

It should be noted that neither the ambient water data nor the boiling water data for the polyimide jacketed fibers show a clear break (or knee) in the their plots. The absence of a break in the static fatigue test results has been reported earlier for Hard Plastic Silica Clad optical fibers[6] and some silicone clad/coated fibers[13]. Final judgment on the long term behavior for these jacketed fibers awaits completion of the tests.

Micro porous structure

The basic science and engineering studies are underway which utilize the micro porous structure to incorporate other materials, which in turn permit the modified sections to interact with the surrounding environment and transmit a modified signal. Preliminary tests have established that complexes such as Rhodamine can be activated by signals traveling in the fiber core after incorporation of the complex in a section of the modified sol-gel clad fiber. The guided waves within the fiber core extend a distance into the cladding. Where the complex is incorporated into the cladding, it can interact with the waves travelling within the core and become activated.

The new studies underway look to explore further these effects to design and construct sensors for various chemical and biochemical moieties. We are looking at several different processing conditions and other variables to create micro pores of different sizes for use with materials of differing dimensions.

Conclusions

A new type of optical fiber has been developed. It is made with all pure silica in both the core and cladding. It has been shown to have high strength properties, good fatigue behavior and good optical properties. With its micro porous structure, it appears to provide sites for other molecules which can be activated by light traveling down the core of the optical fibers. Several different properties have been found, some of which present 'problems' compared to standard optical fibers. Among the latter problems are cleaving, and gripping the fibers for testing or other high stress situations. The micro porous structure can be crushed at high grip pressures needed to tensile test long lengths or it can be consolidated when attempting to fusion splice the fiber with itself or any other all silica fiber. Preliminary spectral experiments have also indicated some unique behavior at wavelengths in the near infrared region of the electromagnetic spectrum. All these aspects are part of the continuing study of these fibers and the push for applications of their unique structure.

Acknowledgments

The author gratefully acknowledges continuing support of this work by the US Army through its Communications and Electronics Command under contracts DAAB07-95-C-D009 and DAAB07-96-C-D612. The author thanks B.G. Bagley, S. DiVita and W. Neuberger for helpful discussions and comments on this work. The author is grateful to S. Johnson, M. Trumbull, and A. Suchorzewski for technical assistance.

References

1. Work performed by CeramOptec, Inc. under a Cooperative Research and Development Agreement (CRADA) on D-shaped fibers with US Army's CECOM at Ft. Monmouth, NJ.
2. Skutnik, B.J. and DiVita, S., "Pure Silica Optical Fibers Utilizing Sol-Gel Techniques", AFCEA Ann. Conference Proc. 1996, **369**-73.
3. Savage, R.O., Fischer, R.J. and DiVita, S., US Patent No. 5,114,738 (1992).

4. MacChesney, J.R., Pinnow, D.A. and van Uitert, L.G., US Patent No. 3,806,224 (1974).
5. Bagley, B.G., Gallagher, P.K., Quinn, W.E. and Amos, L.J., Mat. Res. Soc. Proc. **1984,** *32*, 287.
6. Skutnik, B.J., SPIE **1993**, *1893*, 2 , and references therein.
7. Clarkin, J.P., Skutnik, B.J. and Munsey, B.D., J. Non-Cryst. Solids **1988,** *102*, 106.
8. Wei, T.S. and Skutnik, B.J., J. Non-Cryst. Solids **1988,** *102*, 100.
9. Skutnik, B.J. and Trumbull, M.R., J. Non-Cryst. Solids **1998,** *239*, 210.
10. Norsken 8 meter, 2 coating station, draw tower; Heathway 12 meter. 2/3 coating station, draw tower.
11. Matthewson, M.J., Kurkjian, C.R. and Gulati, S.T., J. Am. Ceram. Soc. **1986,** *69*, 815.
12. Murgatroyd, J.B., J. Soc. Glass Tech. **1944,** *28*, 388.
13. Roberts, D., Cuellar, E., Middleman, L. and Zucker, J., SPIE **1986,** *721*, 28.
14. Skutnik, B.J., Hodge, M.H. and Nath, D.K., in: FOC/LAN '85 Proc., **1985,** 232.
15. Skutnik, B.J., Hodge, M.H. and Clarkin, J.P., SPIE **1987,** *842*, 162.
16. Skutnik, B.J. and Wei, T.S., ibid., 41.
17. Kurkjian, C.R. and Paek, U.C., Appl. Phys. Lett. **1983,** *42*, 251.
18. Skutnik, B.J. and Trumbull, M.R., Mat. Res. Soc. Proc. **1998,** *531*, 169.
19. Kurkjian, C.R., Krause, J.T. and Matthewson, M.J., J. Lightwave Tech. **1989,** 7, 1360.
20. Mathewson, M.J., Rondinella, V.V. and Kurkjian, C.R., SPIE Proc. **1992,** *1791*, 52.
21. Krause, J.T., J. Non-Cryst. Solids **1980,** *38&39*, 497.
22. Cuellar, E., Kennedy, M.T., Roberts, D.R. and Ritter, Jr., J.E., SPIE Proc. **1992,** *1791*, 7.

Chapter 12

Analysis of a Coextrusion Process for Preparing Gradient-Index Polymer Optical Fibers

Yung Chang, Wen-Chang Chen*, Ming-Hsin Wei, and Wen-Chung Wu

Department of Chemical Engineering, National Taiwan University, Taipei, Taiwan 106, Republic of China

A theoretical modeling was introduced on a coextrusion process for preparing gradient-index (GI) polymer optical fibers (POFs). The effects of the mass transfer coefficient (k) between the fiber preform and the purging gas, and the interlayer number (n) on the refractive index distribution (RID) of POFs were investigated. The predicted refractive index distribution (RID) was in a satisfactory agreement with the experimental data reported in the literature. The order of Δn value in the studied systems was $k^*(\sim \infty) > k^*$ (=5) > k^* (~0) for various interlayer numbers. As k closes to a very small or a very large value, the RID deviates significantly from the parabolic RI distribution. The order of the degree of the RI distribution close to a parabolic distribution is five-layer > four-layer > three-layer > two-layer. The result suggests that the greater the layer number in the multilayer coextrusion process, the closer the RID is to a parabolic curve. These results provide a way to monitor the RID of POFs by process design.

Scientific interest in gradient-index (GI) polymer optical fibers (POFs) remains very high in light of their versatile applications in optical communication, imaging, and collimating (*1-6*). New types of high transmission speed of POFs for communication

1: corresponding author

applications have been reported recently. Giaretta et al. developed a GI POF based on perfluorinated (PF) polymer with a transmission speed as high as 11 Gb/s at 1.3 μm (7). A theoretical study on the attenuation limit of PF GI polymer optical fiber (POF) was estimated to be as low as 0.3 dB/km, which is comparable to that of a silica fiber in the near infrared region (1). Koike and his coworkers demonstrated that GI POF can be used as an optical fiber amplifier by doping the fiber with organic dyes (8). Another potential application of GI Polymers is as the essential components of the selfoc lens array (SLA) in fax machines and scanners (9,10).

The refractive index variation inside the GI POF is described by the following two equations:

$$n(r) = n_1 \left[1 - 2\delta(\frac{r}{R})^g \right]^{\frac{1}{2}} \qquad 0 \le r \le R \qquad (1)$$

$$\delta = \frac{n_1 - n_2}{n_1} \qquad (2)$$

Where n_1 and n_2 are the refractive indices of the center axis and the periphery, respectively, R is the fiber radius, and g is the index exponent of the power law. It was found that the bandwidth of a GI POF is maximized when g has the value of 2 (11). That is, the optimal RID is a parabolic curve. Koike et al demonstrated the importance of materials dispersion on the bandwidth of GI POF. They concluded that the g values must be in the range of 1.70-3.0 for achieving high transmission speed (> 1 Gb/s) using a light source with more than 2-nm spectral width (12). For the GI lens applications, keeping both Δn greater than 0.01 and $g \cong 2$ are very important for the imaging applications (9,10). Hence, optimization of the refractive index distribution to a quadratic distribution is very important for GI POF.

In order to achieve a parabolic RID for a POF, both materials system and process design are required to be considered. The differences between the monomer concentration, monomer reactivity, monomer size, monomer density, monomer diffusivity, the properties of the host polymers, and external process design were the driving forces behind producing index gradients inside the fiber. Multilayer coextrusion has been recognized as a potential method for preparing GI POFs. Diffusion of various monomers through each layer and the mass transfer between the fiber preform and the purging nitrogen results in a distribution of refractive index inside the fiber. Based on the purging nitrogen condition on the diffusion zone, it can be classified as the closed process (k =0) and open process (k >0). Our laboratory has done extensive experimental (13-15) and theoretical work (16-18) on a multilayer coextrusion process for preparing GI POF. Furthermore, a general theoretical modeling of the N-layer coextrusion process was developed to predict the effects of the essential parameters on the RID of POF (18). However, the effects of the k values for the high interlayer numbers (four or five layer) on RID of POFs for both the closed and open processes have not been fully explored in the previous study (18).

In this study, an extended work on the theoretical analysis of the multilayer coextrusion process for preparing GI POFs is reported. The effects of the mass transfer coefficient (k) between the fiber preform and the purging nitrogen, and the layer number (n) on the refractive index distribution (RID) of POFs were investigated. The theoretical prediction was compared with the experimental results by us (*13*) and Toyoda et al (*10*). The basic parameters for the present mathematical simulation are shown in Table 1. The effects of the k values on the RID of POFs and the two-five layer coextrusion process were examined.

Coextrusion Process for Preparing GI POF

A similar essential design of the N-layer coextrusion process has been described previously (*18*) and briefly shown here. Figure 1 shows the schematic diagram of the apparatus of the N-layer closed (no nitrogen purging) and open (nitrogen purging) coextrusion process for preparing GI POF. Material tanks (**1~N**) containing a solution of a polymer and monomers were heated at a desired temperature (e.g., 60^0C). Next, the polymer solutions in the tanks (**1~N**) were fed by the gear pumps (**$G_1~G_N$**) to the concentric die (**D**). Here, the refractive indices of the polymer solutions inside the concentric die were arranged in the order of decreasing from the center to the periphery of the die. The order of the refractive index in the N sections is $N_1 > N_2 > N_3 > -------- >N_{N-1}> N_N$. The volume ratios of the polymer solutions from different tanks can be adjusted to obtain the desired refractive index distribution. A N-layer composite monofilament was then extruded out of the orifice of the die and fed into a diffusion zone (**Z**) with a length of L. While the monofilament went through the diffusion zone, the monomer in each layer diffused into each other to produce the refractive index distribution. In the closed extrusion process (a: k=0), the diffusion zone was insulated from the surrounding. In the open process (b:k>0), a hot nitrogen gas was purged through the diffusion zone and thus the mass transfer occurred between the fiber and its periphery. The monofilament was then fed through a hardening zone (**H**) where it was hardened by UV lamps. A polymer fiber with a certain RID was taken up through rolls by a take-up roll (**I**).

Theoretical Analysis

A theoretical modeling based on the above process has been derived previously (18). Briefly, the governing equations and the composition distribution are shown here. Let r and z are the radial distance and the distance measured from the top of the diffusion zone, respectively. R and L are the radius and the length of the diffusion zone. The variation of the mass fraction of monomer M, x, inside the diffusion zone at a steady-state operation can be described by

$$u\frac{\partial x}{\partial z} = D_r(\frac{\partial^2 x}{\partial r^2} + \frac{1}{r}\frac{\partial x}{\partial r}) + D_z\frac{\partial^2 x}{\partial z^2} \qquad (3)$$

Table I. Experimental Conditions[a] Used in the coextrusion process(10,13).

Item	Two-layer	three-layer	Four-layer	five-layer
		Mixture 1 (inner layer)		
PMMA(%)	58	52	52	52
BzMA(%)	28	35	35	35
MMA(%)	14	13	13	13
		Mixture 2 (second layer)		
PMMA(%)	60	50	50	50
BzMA(%)	0	15	15	15
MMA(%)	40	35	35	35
		Mixture 3 (third layer)		
PMMA(%)	-	50	50	50
BzMA(%)	-	-	-	-
MMA(%)	-	50	50	50
		Mixture 4 (fourth layer)		
PMMA(%)	-	-	47	47
BzMA(%)	-	-	-	-
MMA(%)	-	-	40	40
FPMA(%)	-	-	13	13
		Mixture 5 (outer layer)		
PMMA(%)	-	-	-	40
BzMA(%)	-	-	-	-
MMA(%)	-	-	-	18
FPMA(%)	-	-	-	42
R(mm)	0.5	0.59	0.6	0.6
u (cm/min)	254	40	40	40
T($^{\circ}$C)	80	60	60	60
L(cm)	45	45	45	45
Radius ratio(2-layer)	$0.5(R^*_{f1}):1(R^*_{f2})$			
Radius ratio(3-layer)	$0.76(R^*_{f1}):0.96(R^*_{f2}):1(R^*_{f3})$			
Radius ratio(4-layer)	$0.75(R^*_{f1}):0.94(R^*_{f2}):0.98(R^*_{f3}):1(R^*_{f4})$			
Radius ratio(5-layer)	$0.73(R^*_{f1}):0.92(R^*_{f2}):0.96(R^*_{f3}):0.98(R^*_{f4}):1(R^*_{f5})$			

[a]The RI values of PMMA, Poly(BzMA), and Poly(FPMA) used in the present study are 1.490(21), 1.568(21), and 1.425(22), respectively. The densities of PMMA, Poly(BzMA), Poly(FPMA) used in the present study are 1.170(21), 1.179(21), and 1.496(22), respectively.

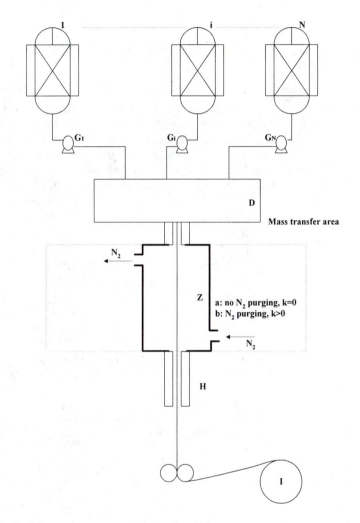

Figure 1. Schematic representation of a N-layer coextrusion process. 1,...,i,...N , material supply tanks; $G_1,...,G_i,...G_N$, gear pumps; D, a concentric die; Z, diffusion zone (a: no N2 purging, b: N2 purging); H, a hardening zone; I, rolls.

where u denotes the extrusion velocity, and D_r and D_z the effective diffusivities of monomers in the r and z directions respectively. The boundary conditions associated with equation (3) are summed as follows :

$$x \text{ is finite at } r = 0 \tag{4a}$$

$$-D_r \frac{\partial x}{\partial r} = kx \text{ at } r = R \tag{4b}$$

where k represents the mass transfer coefficient of monomer, M, between the fiber periphery and the purged nitrogen. For the importance of practical applications, the peclect number (uL/D_z) is large, i.e., the transport of the monomers due to convection motion is more significant than that due to molecular diffusion. Therefore, the last term of equation (3) is much smaller than the term on the left-hand side. In this case, equation (3) can be solved in the following dimensionless mass fraction of monomer M at the outlet of the diffusion zone ($z=L$),

$$x_M^* = 2 \sum_{m=1}^{\infty} \left[\frac{k^* x_{M,N}^* J_0(\lambda_m) + \sum_{i=2}^{N} (x_{M,i-1}^* - x_{M,i}^*) \lambda_m R_{f_{i-1}}^* J_1(\lambda_m R_{f_{i-1}}^*)}{J_0^2(\lambda_m)(k^{*2} + \lambda_m^2)} \right] \times$$

$$\exp\left(-\lambda_m^2 z_M^*\right) J_0\left(\lambda_m r^*\right) \tag{5}$$

In this expression, $r^* = r/R$, J_0 and J_1 are the Bessel functions of the first kind of orders 0 and 1 respectively, λ_m is the positive root of $k^* J_0(\lambda_m) = \lambda_m J_1(\lambda_m)$. And

$$x_M^* = x_M / x_{M,0} \tag{6a}$$

where x_M is the mass fraction of monomer M in the the diffusion zone at z and $x_{M,0}$ is the largest mass fraction of the monomer M in the multi-layer at $z=0$. And

$$z_M^* = \frac{zD_r}{uR^2} \tag{6b}$$

$$k^* = \frac{kR}{D_r} \tag{6c}$$

Extreme cases. Two extreme cases can be recovered directly from the present model : k* is very small (k*→0),and k* is very large (k*→∞). The special case of k*→0 corresponds to a closed co-extrusion. That means there is no mass transfer between the fiber periphery and the purged gas. In this case, equation (5) reduce to equation (7).

$$x_M^* = x_{M,N}^* + \sum_{i=2}^{N}(x_{M,i-1}^* - x_{M,i}^*)R_{f_{i-1}}^{*2} +$$

$$2\sum_{m=1}^{\infty}\left[\frac{\sum_{i=2}^{N}(x_{M,i-1}^* - x_{M,i}^*)R_{f_{i-1}}^* J_1(\lambda_m R_{f_{i-1}}^*)}{\lambda_m J_0^2(\lambda_m)}\right] \times \exp(-\lambda_m^2 z_M^*)J_0(\lambda_m r^*) \qquad (7)$$

where λ_m is the positive root of $J_1(\lambda_m)=0$. The other extreme, $k^* \to \infty$, implies that a perfect permeability at the outer boundary of the diffusion zone. In this case, equation (5) reduce to equation (8).

$$x_M^* = 2\sum_{m=1}^{\infty}\left[\frac{x_{M,N}^*\lambda_m J_1(\lambda_m) + \sum_{i=2}^{N}(x_{M,i-1}^* - x_{M,i}^*)\lambda_m R_{f_{i-1}}^* J_1(\lambda_m R_{f_{i-1}}^*)}{\lambda_m^2 J_1^2(\lambda_m)}\right] \times$$

$$\exp\left(-\lambda_m^2 z_M^*\right)J_0\left(\lambda_m r^*\right) \qquad (8)$$

where λ_m is the positive root of $J_0(\lambda_m)=0$.

The transformation of the fiber composition to refractive index distribution can be performed as below. Lorentz and Lorenz (28,29) suggested that the optical property of a non-absorbing media and its chemical structure correlate through an additive rule. The refractive index of a composite material, for example, can be represented by equation (9), (30).

$$n_d = \sqrt{\frac{1+2\phi}{1-\phi}} \qquad (9)$$

with

$$\phi = \frac{\sum_{M}\frac{n_{d,M}^2 - 1}{n_{d,M}^2 + 2}\frac{x_M}{\rho_M}}{\sum_{M}\frac{x_M}{\rho_M}} \qquad (9a)$$

where $n_{d,M}$ and ρ_M are the refractive index and density of the component M. Therefore, the refractive index of the GI POF by the N-layer extrusion process can be determined by the combination of equation (5) and equation (9). The index exponent (g) of the RI distribution can be obtained by fitting the predicted data with equation (1). Furthermore, the difference between the fiber center and periphery (Δn) can be determined by the difference of the fiber center (n_1) and the periphery (n_2)

Results and Discussion

The present model is justified with the experimental data obtained from the two-layer closed coextrusion process by Ho et al. (*13*) and the three-five layer open coextrusion processes by Toyoda.(*10*). The experimental parameters of these processes are shown in Table 1.

Two-layer Closed Process. Figure 2 shows the variation of RI of a fiber n as a function of its dimensionless radius, r^* at various k^* values for the two-layer coextrusion process. The case of $k^* = 0$ is the closed process while the other two k^* values are the open process. The predicted result at $k^*=0$ by the present model is in satisfactory agreement with the experimental result by Ho et al. (13). The estimated values of Δn and g for $k^* =0$, 5, and ∞ are 0.0178 and 0.95, 0.0179 and 0.96, 0.031 and 1.00, respectively. This result suggests that the Δn value enhances and the RID gets close to a parabolic distribution with increasing the k^* value. The case of $k^*=0$ implies that k=0 from equation (6c) (for a fixed diffusion coefficient D_r). That is, there is no mass transfer between the fiber preform and the periphery. Once the k^* value increases, the mass transfer between the fiber preform and the purging nitrogen increases. That is, the low refractive index monomer MMA evaporating from the fiber preform results in increasing the refractive index difference between the fiber center and periphery. This explains the variation of the RID with various k^* values shown in figure 2. Although the g value increases from 0.95 to 1.00 as k^* increases from 0 to ∞, it is still far away from a parabolic distribution (g =2). There are several approaches to improve the RID of the two-layer coextrusion process: (1) select different polymer/monomer mixture; (2) use a core- shell separation process to cut out the non-parabolic fiber preform after passing through the diffusion zone; and (3) use a higher layer number. From the theoretical estimation, the g values in the range of r^* of 0-0.5 shown in figure 2 are1.41, 1.79 and 1.81 for the k^* value of 0, 5, and ∞, respectively. Hence, the parabolic distribution of GI POFs can be possibly obtained by the core-shell separation design assisted with the theoretical study for the two-layer coextrusion process. An experimental design based on the core-shell separation design has been successfully developed for preparing the GI POFs with a parabolic distribution (14).

Three ~ Five layer IDSE Coextrusion Processes. The three-layer coextrusion process has been extensively studied previously (18). The estimated values of Δn and g in the present study for $k^* =0$, 5, and ∞ are 0.003, 1.01, 0.013 and 2.13, 0.030 and 1.58, respectively. The Δn value is enhanced and the g value gets closer to 2 compared the two-layer coextrusion process shown in the previous section. This result suggests that the addition of a third-layer with a proper selection of polymer mixture can improve the index gradient over the two-layer case. However, the g value deviates from the parabolic distribution as k^* increases from 5 to ∞. This result indicates that an optimum k^* value is required to obtain a parabolic RID. Note that the only monomer present in the outmost layer is MMA as shown in Table 1, with a relatively low RI. Hence, if the mass transfer is too large between the fiber preform and purging

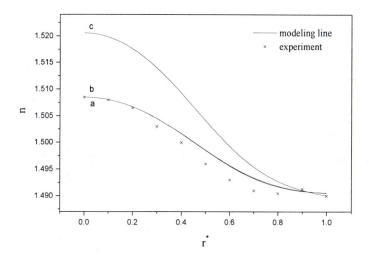

Figure 2. Variation of RI of a fiber n as a function of its dimensionless radius, r^*. at various k^*. Curve a, $k^* \rightarrow 0$; curve b, $k^*=5$; curve c, $k^* \rightarrow \infty$. The experimental conditions are shown in column 1 of Table 1. Parameters used are $z^*_{BzMA}=0.03$, $z^*_{MMA}=0.03$, $R^*_{f1}=0.5$.

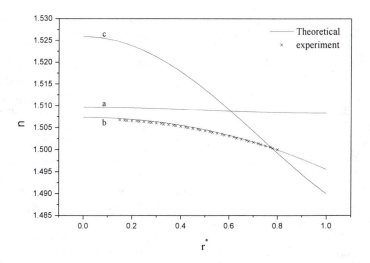

Figure 3. Variation of RI of a fiber n as a function of its dimensionless radius, r^*. at various k^*. Curve a, $k^* \rightarrow 0$; curve b, $k^*=5$; curve c, $k^* \rightarrow \infty$. The experimental conditions are shown in column 3 of Table 1. Parameters used are $z^*_{BzMA}=0.215$, $z^*_{MMA}=0.215$, $R^*_{f1}=0.75$, $R^*_{f2}=0.94$, $R^*_{f3}=0.98$.

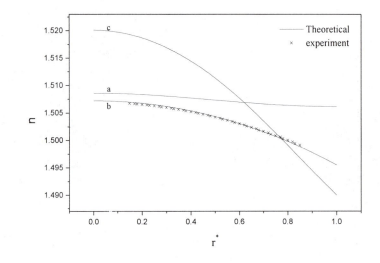

Figure 4. Variation of RI of a fiber n as a function of its dimensionless radius, r^*. at various k^*. Curve a, $k^* \to 0$; curve b, $k^*=5$; curve c, $k^* \to \infty$. The experimental conditions are shown in column 4 of Table 1. Parameters used are $z^*_{BzMA}=0.2$, $z^*_{MMA}=0.2$, $R^*_{f1}=0.73$, $R^*_{f2}=0.92$, $R^*_{f3}=0.96$, $R^*_{f4}=0.98$.

nitrogen, most of the MMA monomer is driven out of the fiber preform but the monomer BzMA with a high boiling point remains. Hence, the g value decreases if k* increases from 5 to ∞.

Figures 3 and 4 show the variation of RI of a fiber n as a function of its dimensionless radius, r* at various k* values for the four and five-layer coextrusion processes, respectively. As presented in the previous publication (18), the predicted results by the present model (at k*=5) are in a satisfactory agreement with the experimental data of Toyoda (15). Here, a further study on the extreme cases (k*→0 and k*→∞) is presented. The estimated values of Δn and g in the four-layer case for k* =0, 5, and ∞ are 0.001 and 1.01, 0.012 and 2.07, 0.036 and 1.55, respectively. For the case of the five-layer case, the Δn and g values for k* =0, 5, and ∞ are 0.002 and 1.01, 0.012 and 2.04, 0.030 and 1.76, respectively. Similar to the three-layer case, as k* closes to a very small or a very large, the RID deviates significantly from the parabolic RI distribution. The g values in the case of k*=5 for the two-five layer coextrusion process are 0.96, 2.13, 2.07, and 2.04, respectively. That is, the more the number of layers, the closer the RID is to a parabolic curve if the proper polymer mixture of each layer is selected. This is because the refractive index gradient of polymer mixture in each layer is easily adjusted as the layer number increases.

Conclusions

A theoretical modeling was successfully developed to predict the multilayer coextrusion process for preparing GI POFs. The order of Δn value increases as the value of mass transfer coefficient (k) increases. As k* becomes very small or very large, the RID deviates significantly from the parabolic RI distribution. The order of the degree of the RI distribution close to a parabolic distribution is five-layer > four-layer > three-layer > two-layer. The result suggests that the greater the layer number in the multilayer coextrusion process, the closer the RID is to a parabolic curve if the proper polymer composition in each layer is selected. These results provide a way to monitor the RID of GI POFs by process design.

References

1. Koike, Y., Ed. Proceeding of Seventh International Conference on Plastic Optical Fibers and Applications, Berlin, October 5-8, **1998**.
2. Kaino, T. in Polymers for Lightwave and Integrated Optics, edited by L. A. Hornak (Marcel Dekker, Inc.: New York, **1992**), Chapter 1.
3. Koike, Y.; Nihei, R. U. S. Patent, 5,593,621, **1997**.
4. Blyler, Jr., L. L., in Ref.1 ,**1998**, A1-A5.
5. Quan, X. Polymer Preprints **1999**, 39, 1279.
6. Garito, A. F.; Wang, J.; Gao, R. Science **1998**, 281, 962.
7. Giaretta, G.; White, W.; Wegmueller, M.; Yelamarty, R. V.; Onishi, T. Proc. OFC/IOOC'99 **1999**, postdeadline papers, PD14-1.
8. A. Tayaga, T. Kobayashi, S. Nakatsuka, E. Nihei, K. Sasaki, and Y. Koike, Jpn. J. Appl. Phys.**1997**, 36, 2705.
9. Yamamoto, T.; Mishina, Y.; Oda, M. U. S. Patent, 4,582,982, **1989**.
10. Toyoda, N.; Mishina, Y.; Murata, R.; Uozu, Y.; Oda, M.; Ishimaru, T. U. S. Patent, 5,390,274, 1995.
12. Ishigure, T.; Nihei, E.; and Koike, Y., Polymer J. **1996**, 28, 272.
13. Ho. B. C.: Chen, J. H.; Chen, W. C.; Yang, S. Y.; Chen, J. J.; Tseng, T. W. Polymer J. **1995**, 27, 310.
14. Chen, W. C. Chen, J. H.; Yang, S. Y.; Cherng, J. Y.; Chang, Y. H.; Ho, B. C. J. Appl. Polym. Sci. **1996**, 60, 1379.
15. Chen, W. C. ; Chen, J. H.; Yang, S. Y.; Chen, J. J.; Chang, Y. H.; Ho, B. C.; Tseng, T. W. ACS Symp. Ser. **1997**, 672, 71.
16. Liu, B.-T.; Chen, W. C.; Hsu, J.-P. Polymer **1999**, 40, 1451.
17. Liu, B.-T.; Hsieh, M.-Y.; Chen,W. C.; Hsu, J.-P.; Polymer J. **1999**, 31 ,233.
18. Chen, W. C.; Chang, Y.; Hsu, J. P, J. Phys. Chem. B **1999**, 103, 7584.
19. Lorentz, H. A. Wied. Ann. Phys. **1880**, 9, 641.
20. Lorenz, L. V.; Wied. Ann. Phys. **1880**, 11, 70.
21. Van Krevelen, D. W. Properties of Polymers, 3rd ed., Elsevier: Amsterdam, The Netherlands, **1990**; Chapters 4 and 10.
22. Tsai, C. C. Ph.D. thesis, National Tsing-Hua University, Hsinchu, Taiwan, 1995.

Chapter 13

The Influence of Mechanical and Climatic Factors on Light Transmission of Polymeric Optical Fibers

P. M. Pakhomov[1], A. I. Zubkov[2], and S. D. Khizhnyak[1]

[1]Physical-Chemistry Department of Tver State University, Sadovyi per. 35,
Tver 170002, Russia
[2]Engineering Center of Polymeric Optical Fibers, Research Institute
of Man-Made Fibers, Tver, Russia

Influence of mechanical deformations (bending and tensile drawing), UV radiation, moisture and elevated temperature on light transmission of polymeric optical fibers (POF) with poly(methyl methacrylate) (PMMA) core and reflecting layer on the base of poly(fluoro acrylate) PFA has been studied by IR-, UV- and ESR spectroscopic, light scattering and optical microscopic techniques. It has been shown that an increase of light losses is caused by an appearance of microcracks under deformation, forming macroradicals and chromophore groups under UV-radiation, a growth of IR absorption on overtone vibrations of OH-groups under damping and also by changing light guide geometry and the destruction processes in it under heating.

Polymeric optical materials are widely used in fiber optics, optical electronics, laser apparatuses and etc. Applying these materials for producing POF is especially promising. The interest to POF is caused by a number of their advantages such as high flexibility, stability to dynamic loading, possibility of creating the fibers with larger diameter, simplicity of their attachment as compared to quartz light guides [1]. It is provided the intensive using POF in medicine, in indicators, reclame and decorative aims.

Safety and effective exploitation of POF needs information about various effects (mechanical, radioactive, dampness, temperature and etc.) on POF transparency. Thus, the aim of this work was to elucidate the influence of such external factors as mechanical deformation, climatic conditions on light transmission of POF.

Experimental

POF with a poly(methyl methacrylate) (PMMA) core coated with a reflecting layer based on poly(fluoro acrylate) (PFA) has been investigated.

$$\left[-CH_2 -C(CH_3) - \right]_n$$
$$| $$
$$CO - O - CH_2 - CF_2 - CF_2H$$

PMMA and PFA were obtained by radical polymerization in bulk ($\overline{M}_w = 8 \cdot 10^4$; $\overline{M}_w / \overline{M}_n = 1.5$). Polymerization was induced by lauryl peroxide and molecular weight was controlled by dodecyl mercaptan addition. The POF samples were prepared by melt extrusion using a circular die. The outer fiber diameter was 560 μm and the reflecting layer was 20 μm thick. PMMA films 50-200 μm thick prepared by casting tetrahydrofuran (THF) solution of PMMA onto a glass support were studied also.

Total optical losses α in POF determined by formula $\alpha = \dfrac{10}{L} \log_{10} \dfrac{I_0}{I} = \dfrac{10}{L} \log_{10} \left(\dfrac{1}{T} \right)$, where I_0 and I are intensities of monochromatic radiation coming into the sample and passing through the sample respectively, T - POF transparency, L- the length of the sample. The measurements were carried out using an instrument described in [2] and equipped with a light source, monochromator and a light detector signal operating as a rule at 640 nm.

The optical losses in POF in visible range (400-800 nm) were measured using a specially designed setup [3], which included integrating sphere of 25 cm in diameter. It allowed to separate the total optical losses into losses due to absorption (α_a) and scattering (α_s) and to measure also α, α_a and α_s under heating POF samples. To estimate losses associated with light scattering instrument [4] has been also used.

The properties of POF samples on tensile drawing under a constant load and on bending were studied. From tensile tests the relationship between static stress σ and optical losses were estimated using the following formula $\alpha = \dfrac{10}{\Delta l} \log_{10} \left(\dfrac{I_0}{I} \right)$, in the case of bending tests the dependence of the optical losses α on curvature R were presented as follows:

$$\alpha = \frac{10}{2\pi\rho} \log_{10} \left(\frac{I_0}{I} \right), \qquad [1]$$

where Δl stands for changes in the length of POF under stress σ; ρ = R+r (r is fiber diameter); I_0 and I are intensities of monochromatic radiation passing through the sample before and after loading [5].

The absorption spectra of PMMA films in UV, visible and IR ranges were measured with spectrophotometers "Specord M 40", "Perkin-Elmer 180", "Perkin-Elmer 1760". The ESR spectra were measured at room temperature with a a "Rubin" spectrometer operated at a wavelength $\lambda=3,2$ cm. The POF samples were placed in a quarts ampoule mounted in the spectrometer cavity. The radical concentration was determined by comparing the first moments of the spectra of the samples with those of the standard $CuCl_2 \cdot 2H_2O$ single crystals with a known concentration of unpaired electrons [6].

The macroradical concentration in the UV-irradiated POF samples was determined in according with [6]. POF samples wound on a 100 mm diameter quartz cylinder were irradiated with the UV light ($\lambda=253,7$ nm) of a BUV-15 lamp arranged coaxially with the cylinder [7]. The irradiation was performed at room temperature either an air or in vacuum. The concentration of various absorbing chemical groups were determined according to Lambert-Bouger-Beer law. Light scattering was studied using an experimental setup based on a Malvern PCS-100 photon correlation spectrometer. A scheme of the setup and the techniques of estimating the scattering particles are described in details in [4,8]. The density and average dimension of microcracks or crazes were estimated by the help of a Leitz Artalux-2 optical microscope. Sorption of the water by polymer films and POF samples have been conducted by exposing them to distilled water directly.

Results and Discussion

The Influence of Mechanical Deformation on POF Light Transmission

During their assembly and operation POF may be subjected to the action of many mechanical stresses (bending, tensile drawing, compression etc.), so it is very important to study the effect of such factors on the optical properties of the light guides. As was shown in [9], one of the causes of poor light transmission is related to light scattering on crazes.

It was found by us [5] that when POF samples are bent optical losses appear to increase as bending curvature decreases according to the parabolic law (Figure 1, curve 1). At a bending curvature higher than R>4 mm, no marked changes in the light transmission of POF were observed. A substantial optical losses were observed only after a further decrease in R. When the samples are relieved of the applied stress, and the free standing fibers take their initial shape, light transmission partially recovers but does not attain the initial value. With decreasing R, an increase in irreversible component of optical losses also follow the parabolic law (Figure 1, curve 2). Hence, total optical losses α determined by formula 1 can be resolved into two components: $\alpha = \alpha_r + \alpha_i$ where α_r and α_i stand for reversible and irreversible losses, respectively.

Reversible component of optical losses is primarily associated with changes in the geometry of POF. For some rays, the condition of total internal reflection does not hold (Figure 2, ray A), and the refracted rays leave the fiber. This results in light emission at the bending regions. A decrease in bending curvature is accompanied by an increase in both the fraction of the rays that leave the light guide and optical losses in POF.

Reversible and irreversible changes in light transmission have been observed in both cases - bending and tensile drawing [5]. We assumed that in both cases (bending and tensile drawing) irreversible optical losses are related mainly to the light scattering on microheterogeneities (crazes) arising in the POF core under mechanical stresses. Actually, when POF samples are drawn or the bending curvature decreases, an increase in the intensity of the light scattering is proportional to the growth in irreversible component of optical losses. The linear dimensions of the scattering defects (in our case, crazes) as estimated from the angular distribution of the indicatrix of the Rayleigh scattering [4,8] were ~ 100 nm.

In the case of tensile drawing reversible component of optical losses is also associated with scattering on crazes, the density of which increases with an increase in the static applied load. However, when the sample is relieved of the applied stress, such crazes are able to "heal" [10] and produce no scattering. Actually, in the stress-relieved samples, we observed a decrease in light scattering and craze density in the POF core.

Direct microscopic observations showed an increase in craze density with a decrease in the bending curvature or with an increase in the strain of POF samples. Figure 3 presents the dependence of craze density on bending curvature. A comparison of Figure 3 and 1 reveals their qualitative similarity. A qualitative comparison of irreversible component of optical losses and craze density reveals their linear dependence. Irreversible component appears to be proportional to the craze density in the POF core.

The data obtained are useful to use in practice to estimate the quality and suitability of POF for exploitation and also to determine the admissible mechanical deformation.

Effect of UV Radiation on Light Transmission in POF

A serious problem hindering the application of POF is the decrease in their light transmission under the action of hard radiation. UV radiation is known [11] to cause degradation of polymeric materials. It was established [12] that increasing the irradiation time leads to a growth of optical losses due to absorption (Figure 4), especially in a short-wavelength (violet) region of the spectrum; POF samples acquire a yellowish color. At that time after irradiation the optical losses due to the light scattering practically did not change.

Studying the dependence of optical losses at a wavelength $\lambda = 650$ nm on irradiation dose showed that an increasing the duration of irradiation causes monotonic and marked rising of optical losses in the sample (Figure 5a, curve 1). After finishing irradiation at small doses (up to 1 hr) light transmission of the sample is restored completely in 200 hr at room temperature (Figure 5b) and yellow coloration disappeared. Upon irradiation at high doses (several hours), irreversible

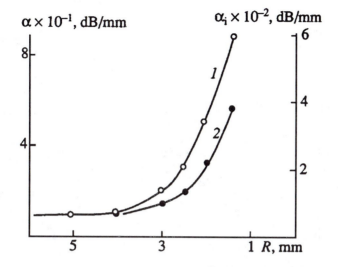

Figure 1. (1) Total and (2) irreversible optical losses in POF versus bending curvature. (Reproduced with permission from reference 5. Copyright 1994).

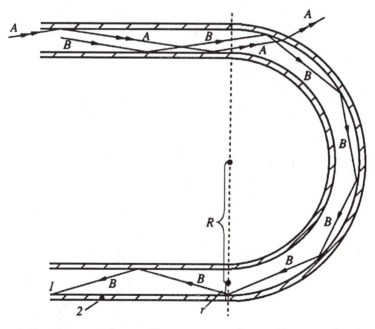

Figure 2. Development of reversible component of optical losses on bending POF. (A) light ray leaving POF, (B) light ray transmitting along POF. (1) POF core, (2) reflecting POF outer layer. R, bending curvature, r, radius of POF. (Reproduced with permission from reference 5. Copyright 1994).

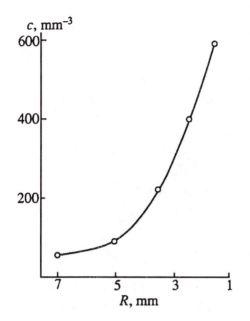

Figure 3. Craze density in POF versus bending curvature. (Reproduced with permission from reference 5. Copyright 1994).

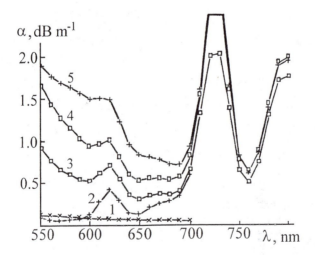

Figure 4. Spectral distribution of optical losses in POF: scattering spectrum of the sample irradiated during 1 hr (1); absorption spectra of the initial sample (2) and the samples irradiated during 20, 30 and 60 min (3, 4, and 5, respectively). (Reproduced with permission from reference 12. Copyright 1993).

changes were observed (a yellowish color and significant optical losses are retained even after prolonged "rest" (Figure 5b)). Irreversible optical losses grow monotonically with increasing doses of irradiation (Figure 5a, curve 2). Thus, for example, after 6 hr of UV irradiation the irreversible component of optical losses reached 2 dB m^{-1}, that limit when POF can be used as a decorative light guide.

Investigation of the mechanism of light transmission in POF using ESR method showed [7] that the photolysis is accompanied by formation and stabilization of radicals (Figure 6). The samples treated for 5 min at room temperature, exhibited an ESR spectrum typical of the peroxide radicals ROO′ (Figure 6a). The concentration of peroxide radicals was about $\sim 10^{16}$ cm^{-3}. Increasing the irradiation time to 10 min led to a growth of the radical concentration to 10^{17} cm^{-3} (Figure 6b). The complex shape of the spectrum suggested that this was due to a superposition of signals from several hydrocarbon radicals. Further increase in the duration of irradiation to 15 and 30 min, and then to 1 and 3 hr resulted in a monotonic growth of the radical concentration to 10^{19} cm^{-3} (Figure 6c). No new lines appeared in the spectrum. And only redistribution of the intensities of the existing individual components took place. As the irradiation dose increased, the shape of the ESR spectrum more and more resembled that of the spectrum of macroradicals of the alkyl type $-CH_2 - \dot{C}$ $(CH_3)(COOCH_3)$. After a 6 hr of UV irradiation, the concentration of radicals reached about $3 \cdot 10^{19}$ cm^{-3}, and the spectrum was typical of terminal alkyl macroradicals.

The storage of the UV-irradiated samples in air at room temperature was accompanied by a monotonic decrease in the concentration of the alkyl radicals to zero (Figure 7a) and no radicals of new types were found to form. During the first 100 hr, the radical concentration dropped about tenfold, and in 25 days it was virtually zero. There is a linear relationship between a value of reversible component of optical losses and the concentration of alkyl radicals (Figure 7b), so that these radicals themselves are the reason of POF optical losses. Alkyl radicals intensively absorb the light at wavelengths λ=200-260 and 425 nm [13]. The high room-temperature stability of these radicals seems to be due to some structural features of the studied samples, determining the mobility of radicals and the oxygen diffusion to the bulk of the polymer. The process studied occurs in PMMA below the glass transition temperature, i.e., under the conditions of "frozen" molecular mobility and hindered oxygen diffusion from air to the radicals formed.

In the case studied, formation and loss of the terminal radicals of the alkyl type during irradiation occurred in phase with the appearance and disappearance of the yellow color. It is therefore natural to assume that these alkyl type radicals are responsible for reversible variations in the coloration and light transmission of POF.

The optical spectrum of PMMA, especially in samples irradiated for more than 1 hr, shows the absorption due to the accumulation of chromophore groups C=C and C=O [7]. This is manifested primarily in the absorption bands with peaks at wavelengths of 195, 218 and 280 nm (Figure 8). The band with λ_{max} = 195 nm is attributed to isolated C=C bonds. The absorption coefficient of the C=C bonds is large, and even a small concentration of these groups yields a noticeable absorption band in this region (Figure 8, spectrum 6). The band with λ_{max} = 218 nm is assigned

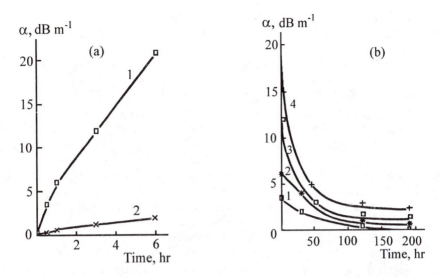

Figure 5. (a) dependence of optical losses in POF (λ = 650 nm) on duration of UV irradiation: 1 - total optical losses; 2 - irreversible component of optical losses. (b) recovering light transmission in POF irradiated for 30 min (1), 1 (2), 3 (3), and 6 hr (4) at λ = 650 nm. (Reproduced with permission from reference 12. Copyright 1993).

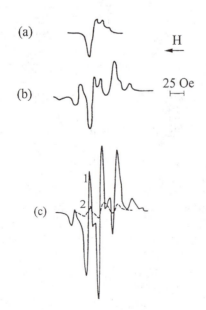

Figure 6. ESR spectra of polymeric optical fibers with PMMA core UV-irradiated for 5 (a), 10 (b), and 15 min (c) measured immediately after the irradiation (1) and after 140 hr storage (2). (Reproduced with permission from reference 7. Copyright 1992).

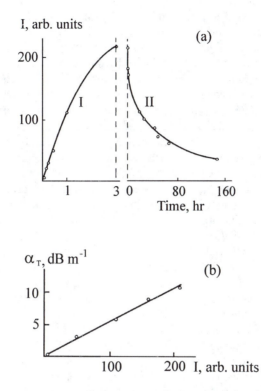

Figure 7. (a) intensity of the ESR signal characterizing the content of alkyl radicals in PMMA versus the duration of UV irradiation (I) and storage (II). (b) the reversible component of optical losses in POF versus the content of alkyl radicals. (Reproduced with permission from reference 7. Copyright 1992).

to ester groups of the initial PMMA. This region also contains some contribution from the C=C bonds which are formed under the photolysis in such molecular chain structure as ~CH=C(CH$_3$)(CH$_2$)~ and ~CH=C(CH$_3$)(COOCH$_3$). The absorption band with λ_{max} = 280 nm is attributed to C=O groups formed under the action of UV radiation. These can be keto- and ether groups entering into the macromolecular chain, and acidic and aldehyde groups at the chain ends.

Analysis of the UV spectra (Figure 8) shows that the photolysis of the PMMA films in air is accompanied by an intense growth of the absorption band having a peak at 195 nm. This may be evidence for a fast accumulation of chromophore groups C=C in the polymer. The presence of the band at 195 nm in the initial (nonirradiated) polymer is likely to indicate the presence of a residual monomer. Indeed, keeping the sample in vacuum (p = $2 \cdot 10^{-1}$ mm Hg) produced a noticeable decrease in the intensity of this band.

The character of accumulation of the C=C and C=O groups in PMMA with increasing irradiation time is illustrated in Figure 9. During the post-irradiation storage, the concentration of the C=C and C=O groups remained virtually unchanged, as did the yellowish color of the samples. The intensity of coloration increased with the duration of irradiation.

The dependence of the irreversible component of the optical losses in POF on the number of photolysis-induced chromophore groups is shown in Figure 10. The linear relationship between the irreversible optical losses and the concentration of C=C and C=O groups in PMMA may be an indication of the decisive role of the C=C and C=O groups in the core materials of POF in the mechanism of irreversible losses.

Thus, at UV irradiation of short duration (to 0.5 – 1 hr), the optical losses are determined by the processes of formation and loss of macroradicals in the POF core. There is a linear relationship between the reversible losses and the concentration of macroradicals (predominantly, of the alkyl type). Further increase in the irradiation time leads to a considerable accumulation of double bonds, C=C and C=O, in PMMA, with a resulting intense absorption in the UV and visible ranges. These processes determine the irreversible component of optical losses in POF.

Effect of Water on Light Transmission in POF

PMMA widely used as a core of POF can absorb up to 2% of water at ambient temperature and pressure [14]. The water thus sorbed, although small in amount, significantly affects the mechanical, optical, and some other properties of polymer. Polymeric light guides are employed, as a rule, in the visible and near-IR (400 – 1300 nm) ranges, i.e., in the region where overtones and combination frequencies of the OH vibrations manifested.

It is known [15,16] that the presence of water in POF may considerably reduce their optical transmission in visible and near-IR regions. The results of the spectroscopic measurements showed (Figure 11) [16] that sorption of water by PMMA films produces a drastic increase of the absorption bands at 7028, 5243, 3650, 3560, and 1640 cm^{-1}, corresponding to stretching and bending vibrations of OH

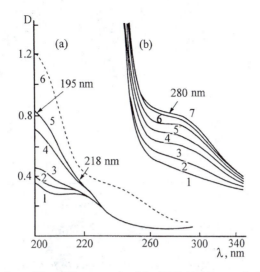

Figure 8. Variation of the UV spectrum of the PMMA films with thicknesses of 5 (a) and 130 μm (b) with the duration of UV irradiation. (a) initial spectrum (1), 0.5 (2), 1.5 (3), 4.0 (4), 5.5 hr (5), spectrum of a 1 mm thick MMA layer (6). (b) initial spectrum (1), 1.0 (2), 2.0 (3), 3.0 (4), 4.0 (5), 5.0 (6), and 6.0 hr (7). (Reproduced with permission from reference 7. Copyright 1992).

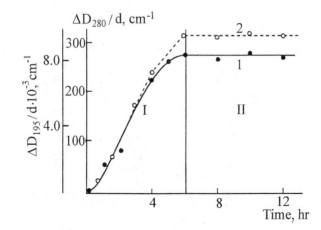

Figure 9. Variation of the extinction of the UV absorption at 195 (1) and 280 nm (2) with the duration of UV irradiation (I) and storage (II). (Reproduced with permission from reference 7. Copyright 1992).

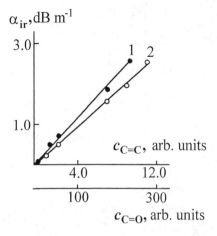

Figure 10. Irreversible component of the optical losses in POF versus the relative contents of C=C (1) and C=O groups (2). (Reproduced with permission from reference 7. Copyright 1992).

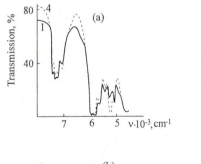

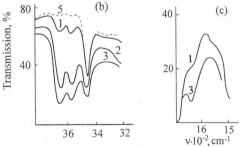

Figure 11. IR spectra of 1 cm thick (a) and 250 μm thick (b, c) PMMA samples in the initial state (1); after exposure to water at 293 K for 10 (2), 20 min (3), and 48 hr (4); and after drying at 353 K over P_2O_5 in vacuum for 5 hr (5). (Reproduced with permission from reference 16. Copyright 1992).

groups [2]. As the duration of polymer exposure to water increased, the amount of the absorbed water and, hence, the intensity of the IR absorption bands corresponding to OH vibrations grew. The desorption of water from polymer led to a virtually complete recovery of the initial band intensities. Note that the IR spectrum of the initial film (Figure 11b, spectrum 1) also contains the absorption bands due to the vibrations of the OH groups representing the equilibrium atmospheric moisture. The experiment on the drying films in the Fisher's gun over P_2O_5 with vacuum-pumping showed a further decrease in the intensity of the above absorption bands (Figure 11b, spectrum 5).

The sorption of water in PMMA significantly depends on the sample thickness and the temperature of water. Since water sorption in the polymer occurs by diffusion, the higher sample thickness implies the slower attainment of the equilibrium water content. Increasing the water temperature resulted in higher sorption by the polymer, as confirmed by the growing intensities of the IR absorption bands due to the OH stretching vibrations. The amount of water absorbed by PMMA increased monotonically with the temperature until reaching the glass temperature of the polymer. Elevating the water temperature increases the molecular flexibility of PMMA chains and the mobility of the molecules of water. As a result, the ability of the latter to penetrate the polymer grows. The exposure of the films to boiling water resulted in their contraction and rendered them turbid and brittle. This can be explained by proximity of the PMMA glass transition temperature to the boiling point of water, and the possibility of hydrolysis of the polymer chains at the ester group [14] and also by the increasing ability of the water molecules having penetrated into the polymer to cluster, which will also lead to the degradation of the material.

The experiment on water absorption by POF revealed (Figure 12) that exposure to water also considerably affects its transmission. The water penetrating into POF is manifested in the form of intense bands peaked at the wavelengths of approximately 920 and 950 nm, which can be assigned to the second overtones of asymmetric and symmetric OH stretching vibrations, respectively. In addition to this, a significant water-induced absorption band is observed in the region of 700 – 850 nm, corresponding to the third-overtone vibrations of the OH groups. Finally, in all cases, water sorption results in an increase in the general background of the optical losses. This phenomenon may arise from an increase in the Rayleigh scattering of light by water molecules having diffused into the polymer, this is known to be one of the major factors responsible for the optical losses in POF [17].

Thus, water absorbed by POF core leads to considerable decreasing its light transmission. It was also established [18] that not only water but another liquids have essential influence on PMMA light transmission.

Effect of Elevated Temperature on POF Light Transmission

Elevated temperature is one of the main factor having influence on POF exploitation ability. The results of conducted investigations showed [19] that elevated temperature leads to a noticeable increase of the total optical losses in all visible range (Figure 13, curve 1). The growth of optical losses is determined by the

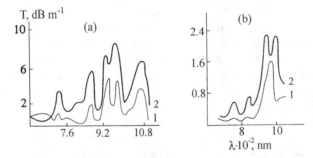

Figure 12. Spectra of total losses of PMMA-PFA (a) and PMMA$_{D8}$-PFA (b) optical fibers in the initial state (1) and after exposure for 17 hr to water (2). (Reproduced with permission from reference 16. Copyright 1992).

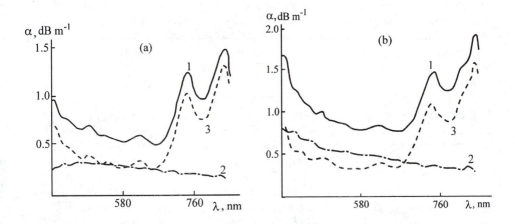

Figure 13. Spectrum of total losses at 20 °C (a) and 75 °C (b). 1 - total optical losses (α), 2 - absorption (αₐ), 3 - light scattering (αₛ). (Reproduced with permission from reference 19. Copyright 1996).

increasing optical losses due to the scattering (curve 2) and light transmission (curve 3).

Detailed analysis of temperature dependence of optical losses (Figure 14, curve 1) established that light guide can be exploited normally up to 70 ^{0}C, as in the 20 – 70 ^{0}C region the drastic decreasing the light transmission has not been observed. Further increase of the temperature causes an intense growth of optical losses associated primarily with PMMA transition in high elastic state that results in irreversible changes of light guide geometry. The change of POF geometrical sizes (light guide's sagging subjected to the action of own weight) at the temperature higher than 70 ^{0}C was observed visually. At 100 ^{0}C total optical losses reached 1,6 dB/m, the limit when POF hardly can be used even as a decorative light guide.

As is seen from Figure 14 the total optical losses sums up from losses due to absorption (curve 2) and scattering (curve 3). Up to 70 ^{0}C a part of both components is practically equal. With further increasing the temperature the absorption part in total optic losses started to predominate over the scattering part. The study of light transmission mechanism in POF showed [9, 20-21] that the vibrations of groups C=O and C-H in PMMA at 1730, 2950 and 1990 cm $^{-1}$ give a general contribution in characteristic absorption of POF with PMMA core. With increasing the temperature an increase of half-width of IR absorption band at indicated wavenumbers has been observed [19], which is connected with the growth of the molecular mobility intensity. Especially noticeable changing the half-width of the bands occurred at temperature higher than 70 ^{0}C. Widening the considered bands leads inevitably to increasing the absorption in overtone vibrations region in which POF is used. Moreover, the intensive increasing the molecular mobility under the polymer transition in high elastic state promotes the crystallization, that causes the heterogeneity of the sample and decreasing the light transmission due to the scattering on crystalline regions. It was found [19] also that at the temperature higher than 70 ^{0}C the marked accumulation of oxygen-containing groups C=O as a result of thermodestruction is observed. This destruction results in as the growth of absorption in visible region due to increasing the intensity of overtone vibrations of C=O groups as due to the increasing the intensity of absorption of chromophore C=O groups from the side of UV range [7]. Moreover, the destruction process in the polymer is accompanied by the appearance and growth of the microcrack number [10,22] that leads to increasing the scattering in POF core material.

Thus, the heating POF above 70 ^{0}C induced the intense growth of optical losses due to the increasing the molecular mobility and geometric distortion of the light guides, and due to the destruction processes.

Conclusions

As a result of investigating POF with PMMA core and a reflecting layer on the base of PFA the mechanism of optical losses has been elucidated relying on the data from ESR, optical spectroscopy and microscopy, light scattering. This approach can be applied in practice to estimate the quality and stability of POF under certain

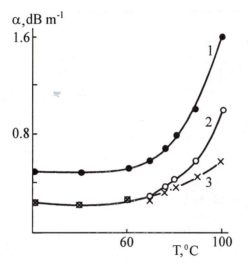

Figure 14. Temperature dependence of total optical losses in POF at λ = 640 nm. Duration of POF at indicated temperature is 5 min. 1 - total optical losses (α), 2 - absorption (α_a), 3 - scattering (α_s). (Reproduced with permission from reference 19. Copyright 1996).

climatic factors and operating conditions and allows also to select optimal conditions for assembling operating POF.

References:

1. Kaino, T. J. Sen-i-gakkaishi. **1986**, vol. 42, no.4, p. 21.
2. Gilbert, A.S.; Petric, R.A.; Phillips, D.W. J. Appl. Polym. Sci. **1977**, vol. 21, no.2, p. 319.
3. Maryukov, M.A. J. Kvantovaya Elektron. **1988**, vol. 15, no.5, p. 1080.
4. Koike, Y.; Tanio, N.; Ohtsuku, Y. Macromolecules. **1989**, vol. 22, no.3, p. 1367.
5. Pakhomov, P.M.; Zubkov, A.I.; Khizhnyak, S.D. J. Vysokomolekulayrnye Soedineniya. **1994**, vol. 36B, no.8. p. 1379.
6. Lebedev, Ya.S. J. Uspekhi Khim. **1968**, vol. 37, no5, p. 934.
7. Pakhomov, P.M.; Fenin, V.A.; Levin, V.M.; Chegolya, A.S. J. Vysokomolekulayrnye Soedineniya. **1992**, vol. 34A, no.11, p. 146.
8. Pakhomov, P.M.; Zubkov, A.I.; Khizhnyak, S.D.; Baran, A.M.; Levin, V.M. J. Vysokomolekulayrnye Soedineniya. **1998**, vol. 40A, no.9, p. 1451.
9. Emslie, Ch. J. Mater. Sci. **1988**, vol. 7, no.7, p. 2281.
10. Volynskii, A.L.; Bakeev, N.F. Vysokodispersnoe Orientirovannoe Sostoyanie; Khimiya: Moskow, 1984.
11. Ranby, B; Rabek, J.F. Photodegradation, Photooxidation, and Photostabilization of Polymers; London, 1975, p.647.
12. Pakhomov, P.M.; Maryukov, M.A.; Levin, V.M. Chegolya, A.S. J. of Applied Spectroscopy (Belorussia). **1993**, vol. 59, no1-2, p. 92.
13. Zhdanov, G.S.; Khamidova, L.G.; Milinchuk, V.K. J. Khim. Vys. Energ. **1983**, vol. 17, no.1, p. 47.
14. Smith, L.S.; Scheniz, V. Polymer. **1988**, vol. 29, no.10, p. 1871.
15. Avakian, P.; Hsu, W.V.; Meakin, P.; Snyder, H.L. J. Polym. Sci., Polym. Phys. Ed. **1984**, vol. 22, no.9, p. 1607.
16. Pakhomov, P.M.; Kropotova, E.O.; Zubkov, A.I.; Levin, V.M.; Chegolya, A.S. J. Vysokomolekulayrnye Soedineniya. **1992**, vol. 34A, no.11, p. 139.
17. Takemuro, T. Polym. Appl. **1987**, vol.36, no.3, p.135.
18. Pakhomov, P.M.; Khizhnyak, S.D.; Belyakova, T.I. J. Vysokomolekulayrnye Soedineniya. **1995**, vol. 37A, no.2, p. 230.
19. Khizhnyak, S.D.; Pakhomov, P.M.; Zubkov, A.I. J. Vysokomolekulayrnye Soedineniya. **1996**, vol. 38B, no.9, p. 1623.
20. Kaino, T. Kobunshi Rombunshi. **1985**. vol. 42, no.4, p. 257.
21. Groh, W. Macromol. Chem. **1988**. B.189, no.12, p. 2861.
22. Regel', V.R.; Slutsker, A.I.; Tomashevskii, E.E. Kineticheskaya Priroda Prochnosti Tverdykh Tel; Nauka: Moscow, 1974. p.117.

Chapter 14

Development of a Fiber Optic pH Sensor for On-Line Control

M. Janowiak, H. Huang, S. Chang, and L. H. Garcia-Rubio

Department of Chemical Engineering, University of South Florida, Tampa, FL 33620

Fiber optic chemical sensors is a field of growing importance with diversified applications in oceanography, chemical process, clinical diagnosis, and environmental monitoring. The goal of this study is to develop the technology to support specific analytical reagents on polymeric membranes for fiber optic based pH sensors. For on-line process control applications, pH sensors must have a fast response, high sensitivity, and be able to withstand relatively high temperatures. A three layered method has been developed to grow a hydrophilic polymer matrix from a fiber optic and to incorporate specific pH indicator dyes. This chapter reports on the synthesis conditions of the polymer matrix to attain light transparency, fast response, and thermal and mechanical stability. The implications of the matrix behavior, relative to the interpretation of the signal and to the flexibility of the system for its use with other indicators, are also presented and discussed. An in-depth look at the interpretation model and the accuracy and precision attainable is taken.

Fiber-optic based chemical sensors is an area of growing importance with diversified applications in oceanography, chemical process, clinical diagnosis, and environmental monitoring. Within these applications, pH sensors play an important role in many biomedical and industrial applications such as blood-gas analysis and process control.

Fiber optic based pH sensors, or pH optrodes, have been shown to have the necessary sensitivity for continuously monitoring changes in the chemical composition of a large variety of process environments including slurries, blood, and low ionic strength fluids in nuclear plants [1-8]. For on-line process control applications in the chemical industry, pH sensors must have a fast response, cover a broad range of pH values with high sensitivity, and should be able to withstand relatively high temperatures [4]. Therefore, it is important to develop a transparent, flexible support matrix that enables the incorporation of a large variety of indicator dyes and that has the necessary thermal and mechanical stability for continuous operation. A three layered synthesis method has been developed for this purpose (Fig. 1-4). This method covalently attaches a hydrophilic polymer matrix onto a fiber optic and enables the incorporation of specific pH indicator dyes. Wherein, the synthesis conditions to attain high light transparency and adequate thermal and mechanical stability are reported. The implications of the matrix behavior relative to the interpretation of the signal, and to the flexibility of the system for its use with other indicators, is also presented and discussed.

Experimental

A schematic of the necessary synthesis steps to develop the three-layered pH sensor is depicted in Figures 2-4. A diagram of the proposed sensor is shown below in Figure 1. The experimental details are given in reference [4].

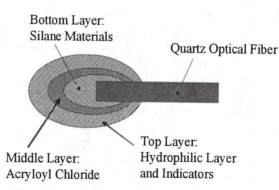

Figure 1. Proposed three-layered design of the optrode.

Surface Silanization

A 0.28 mmol/L aminopropyltriethoxysilane aqueous solution was prepared by mixing 1.3 ml 3-aminopropyltriethoxysilane with 20 ml deionized water in a small bottle with a cap, then adjusted to pH=2 with HCl and heated to 75°C in an oven for 30 minutes. A glass slide, 1 x 3.6cm, was submerged in the solution and maintained at 75°C for 50 minutes. Then, the slide was removed from the bottle and dried at 165°C for 3 hours under an argon purge. This is depicted in Figure 2. IR(KBr): 3500 cm^{-1} (s), 3160 cm^{-1}(s), 1640 cm^{-1}(m), 1590 cm^{-1}(m), 1140 cm^{-1} (s).

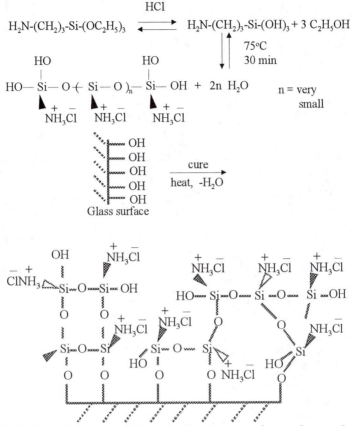

Figure 2. Surface silanization.(Adapted with permission from reference 5. Copyright 1996.)

Activation of Amino Glass

The above slide was treated with pyridine for 4 hours and then reacted with acryloyl chloride at room temperature in the dark for 3 hours, shown below in Figure 3. IR(KBr): 3500 cm^{-1}(s), 3160 cm^{-1}(s), 1720 cm^{-1}(m), 1640 cm^{-1}(m), 1590 cm^{-1}(m), 1140 cm^{-1}(s).

Polymerization

A beaker containing a mixture of 2.1745 g (0.3 mol) of acrylamide (1), 0.3658g (0.0024 mol) of N,N-methylene bisacrylamide (2), 0.5 ml (0.0046 mol) of 4-

vinylpyridine(3), 0.9703 g (0.004 mol) of ammonium persulfate(4), 20 ml of ethanol and 20 ml of deionized water was stirred on a hot plate magnetic stirrer. An activated glass slide was put in the beaker, which was sealed with Parafilm and heated to a reaction temperature of 78°C for 20 minutes. A polymeric gel was obtained, Figure 4.

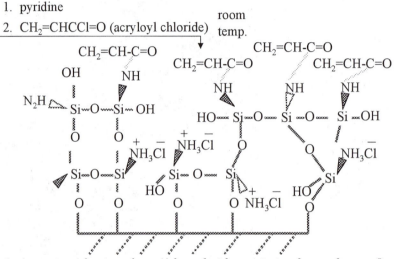

Figure 3. Activation of amino glass. (Adapted with permission from reference 5. Copyright 1996.)

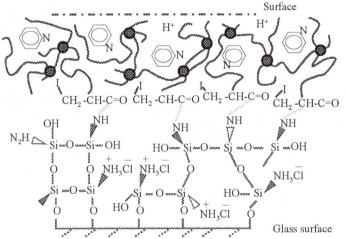

Figure 4. Polymerization. (Adapted with permission from reference 5. Copyright 1996.)

Results and Discussion

Synthesis

Surface silanization is the key to successfully attaching the hydrophilic film onto the glass surface [6]. Figure 2 shows the suggested reaction mechanism. 3-Aminopropyltriethoxysilane is cleaved by acid to form free silanol(Si-OH) groups. Because the structure of silanol is similar to that of glass(silicon dioxide), silanols tend to deposit on the glass surface. It appears that the reaction initially goes in the right hand direction to produce silanols and ethanol. Ethanol is soluble in water and also evaporates into the atmosphere at a reaction temperature of 75°C. This is the driving force for the cleavage reaction to go to completion. On the other hand, as the silanol concentration increases, another equilibrium is established. The silanols start to condense between them to form oligomers. An increase of solution viscosity suggests that polymers are forming in the solution. This reaction is prevented from going to completion by the water in the solution.

HCl was selected as the catalyst for the cleavage reaction and condensation reaction, and as a protective agent for amino groups, by preventing them from participating in the chelation reaction. After completing the surface silanization step, pyridine is used to restore the amine groups for subsequent reactions. The reaction temperature for the curing reaction was established at 165°C and the reaction time to be between 2.5 and 3.5 hrs. These conditions were validated with IR spectroscopy.

The concentration of 3-Aminopropyltriethoxysilane solution can affect the thickness of the layer and therefore the residual hydroxyl groups in this layer. The hydroxyl group is highly hydrophilic and would present potential weak points in the layer, which may cause this layer to peel off from the glass when immersed in water at high temperatures. From the IR intensities of O-H and N-H stretching bands the concentration of the 3-Aminopropyltriethoxysilane was established at 6.5% (by volume). Scanning electron micrographs show that the silane layer is a multilayer of approximately 3~4 μm. Our experiments indicated that the three dimensional crosslinking between silanols significantly enhances the water stability of the silane layer in agreement with the work of Plueddemann [7].

The activation of amino groups was carried out in pyridine to remove HCl and incorporate vinyl groups(acryloyl chloride) into this layer. The reactions were conducted at room temperature and in the dark to protect the double bonds from reaction. In the IR spectra, the characteristic peaks at 1720 cm^{-1} and 1640 cm^{-1} can be assigned to carbonyls and carbon-carbon double bonds, respectively, thus providing evidence for the existence of vinyl groups.

In the polymerization step, the ratio of N,N-methylene bisacrylamide to acrylamide can be adjusted to provide different degrees of crosslinking, therefore affecting the response time of the sensor. Generally, a higher degree of crosslinking is accompanied by a longer response time. The optimization of this step is still under investigation. Ammonium persulfate was used as an initiator, because it does not absorb in the UV-visible range and has good water solubility. Solution polymerization is used for glass slides whereas photo-polymerization is used for fibers. The advantage of photopolymerization is that it provides a better control of the thickness of the polymer film. This is accomplished by controlling the light source intensity and reaction time. However, the aromatic ring of photochemical initiators yields a wide absorption in UV-visible range, which introduces a chromophore impurity. The photopolymerization reaction is still under investigation

Solution polymerization yields adequate films, as depicted in Figure 5, where the UV-visible spectra show that the three layer polymeric thin film has no absorption between 250~800 nm. Spectrum A is of the glass slide, and spectrum B is the three-layered matrix on the glass slide. It has very good light transparency and has a wide spectral window to incorporate different organic indicators.

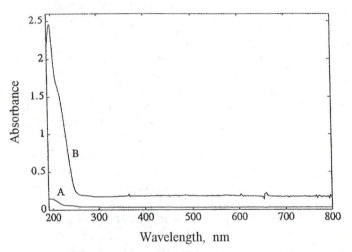

Figure 5. UV-Vis spectrum of a three-layered support matrix. Spectrum A is of the glass slide, and spectrum B is the three-layered matrix on the glass slide.

The polymeric matrix proved to be very stable in water. It does not peel off from the glass surface after being soaked in water for eight months, nor does it after being put in an oven to dry. Furthermore, the polymer membrane shows no evidence of peeling off even after being autoclaved at 120°C in hot water.

Pyridine was chosen as a very simple indicator to incorporate into the polymer matrix. Figure 6 was recorded for a polymer film with vinylpyridine at different pH in water. The spectra indicate that pyridine is only sensitive in a range of pH=2.0~4.0. Pyridine is a base and therefore a good indicator for a low pH range. The potential for the use of multiple indicators is shown in Figure 7, where the spectra of phenol red as function of pH are shown. Note that, although it appears that there are no spectral windows for the combined use of pyridine and phenol red within the same membrane, a fiber bundle can be easily designed. In this case each fiber will have a single indicator and it will be utilized for the appropriate pH range.

This approach will maximize the sensitivity and provide redundancy for improved statistics.

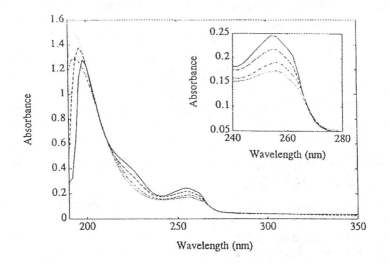

Figure 6. UV-Vis spectra of vinyl pyridine in the three-layered support matrix. (Reproduced with permission from reference 5. Copyright 1996.)

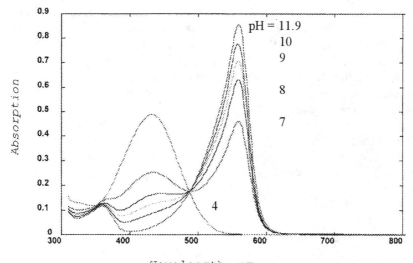

Figure 7. UV-Vis spectra of phenol red-based pH sensitive polymer sensor.

Interpretation of the Detector Response

Accurate interpretation of the spectra of the pH sensitive membranes requires the understanding of the effects of temperature, ionic strength, and buffering capacity of the indicator dyes on the physical properties of the membranes. In particular, swelling and scattering effects due to changes in the optical properties of the sensor membranes can considerably bias the interpretation of the spectra in terms of pH. A comprehensive model for the interpretation of the spectra from pH optrodes has resulted in the following interpretation equation [8],

$$pH = pk_{ao} + P\log(Cond) + M\left(\frac{1}{T}\right) + \log\left(\frac{X_2}{X_1}\right) \tag{1}$$

X_1 and X_2 correspond to the fraction of protonated and dissociated forms of the indicator. That is, the ratio $X_2/X_1 = [A^-]/[HA]$. The term *Cond* corresponds to the conductivity of the solution, and M and P are adjustable or calibration parameters.

The effects on the pH measurement can be broken up into two categories, physical-chemical effects and optical effects. The physical-chemical effects of interest are temperature, ionic strength, and buffering capacity of the indicator dyes.

Buffering Capacity

Immobilizing the indicator dyes in the membrane causes some bias in the pH measurement. The locally high concentration of dye leads to charged groups in the matrix. The swelling of the membrane, which should be optimized to maintain the matrix stability, can reduce this charge. The ionic strength of the solution plays a role in the buffering capacity of the dye. At low ionic strength, the charged groups tend to neutralize themselves with the hydrogen ions, causing a buffering effect. If the concentration of positive ions in the solution is higher, this effect is lessened. Due to the increases concentration of positive ions attracted by the negative charges in the matrix, the pH in the membrane is higher than that of the solution. Since the buffering effect causes a shift in the dissociation equilibrium, it is necessary to know the dye concentration to quantify this effect. The dye concentration should be optimized to minimize these effects while optimizing the signal-to-noise ratio of the sensor.

Ionic Strength

The ionic strength of the solution effects the measured pH through the equilibrium constant of the dye. At low ionic strength, the pK_a of the indicator is no longer constant [8], seen in Figure 8. As the concentration of ions increases, the negative charges in the matrix are shielded causing less of an effect on the equilibrium constant of the dye.

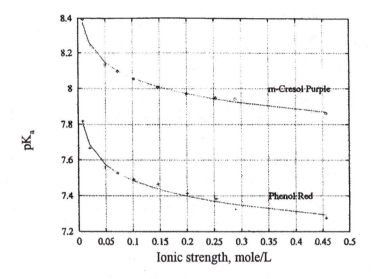

Figure 8. Change in pKₐ as a function of ionic strength.

To account for the effect of ionic strength, the following kinetic model has been proposed [8],

$$X_2 + pS \underset{k_d}{\overset{k_c}{\rightleftharpoons}} X_2^* + pS^*$$

$$X_2^* + H^+ \underset{k_b}{\overset{k_a}{\rightleftharpoons}} X_1$$

$$X_2 + H^+ \underset{k_b}{\overset{k_a}{\rightleftharpoons}} X_1$$

where the salt, S, in the solution collides with reactant X_2^- bring it and the salt to excited states, X_2^{-*} and S^* respectively. The collision frequency required to provide the energy for reaction is reflected in the parameter p. The excited reactant along with the unexcited reactant interact with the hydrogen ions. Essentially, this proposes two different rates of reaction. $[S/S^*]$ is directly proportional to the ionic strength, giving a relationship of the following form, $[S/S^*] = -k'[I]$. Combining the equilibrium constants, $K_a = \dfrac{K_a}{K(k')^p} = \dfrac{[X_2][H^+][I]^p}{[X_1]}$. Taking the logarithm, the equation becomes, $pK_a = pH - \log \dfrac{[X_2]}{[X_1]} - p \log I$. In terms of the thermodynamic equilibrium constant, $pK_a = pK'_{ao} + P \log I$, where P=-p. This model has been tested and adequately represents the behavior of a number of indicator dyes. A more easily measured quantity than ionic strength is conductivity. For this reason, the model has been expressed in terms of conductivity. The conductivity of a solution is defined as, $Cond = F \sum_i |Z_i| u_i C_i$, where F is Faraday's constant, C is the concentration of ions, Z is the charge, and u is the mobility, defined as, $u_i = \dfrac{|Z_i| e}{6 \pi \eta r_i}$, where e is the charge of one electron, η is the viscosity of the medium and r_i the radius. Using the definition of ionic strength, $I = \dfrac{1}{2} \sum_i C_i Z_i^2$, combining with the conductivity, and assuming the ions have similar hydrodynamic radii, $Cond = \dfrac{Fe}{3 \pi \eta r} I$. Taking the logarithm of this expression and replacing in the model for the equilibrium constant, the following equation is obtained [8],

$$pK_a = pK'_{ao} + P\left(-\log \frac{Fe}{3\pi\eta r} + \log Cond\right). \tag{2}$$

Temperature

The temperature dependence of the dissociation constant can be represented by an Arrhenius relationship, $k(T) = A \exp \dfrac{-E}{RT}$, where A is the pre-exponential factor, E is the activation energy, R the gas constant, and T temperature. In terms of the equilibrium constant, $K = \dfrac{k_1}{k_2} = \dfrac{A_1}{A_2} \exp \dfrac{(E_2 - E_1)}{RT}$. k_1 and k_2 are the dissociation constants of the forward and reverse reactions of the indicator dye. Combining terms, the effect of temperature on pK_a has the following form [8],

$$pK_a = A' + M \, \frac{1}{T}. \tag{3}$$

A' is related to the log of the ratio of the pre-exponential factors, and M is related to the difference in the activation of the forward and reverse reactions.

Combining the terms for temperature and ionic strength, equations 2 and 3, the following equation is obtained [8],

$$pK_a = pK_{ao} + P \log(Cond) + M\left(\frac{1}{T}\right) \tag{4}$$

where pK_{ao}, P, and M are estimated calibration parameters. This equation accounts for the pertinent physical-chemical effects on the pH measurement.

The concentrations of the protonated and dissociated forms of the indicator dye are related to absorbance by the Beer-Lambert law, $\tau(\lambda) = \ell \sum_{i=1}^{N} C_i \varepsilon_i(\lambda)$, where τ is the optical density, ℓ the path length, C the concentration, ε the extinction coefficient, and i=1 and i=2 represent $[X_2]$ and $[X_1]$, respectively. The optical corrections that need to be made are for the difference in refractive index between the solution and the membrane and scattering.

Refractive Index Correction

Due to the difference in refractive index between the solution and the membrane, the spectrum is shifted. This can be corrected relative to a reference, the wavelength in vacuum, λ_o, for example, $\lambda_o = \lambda_1 n_1 = \lambda_2 n_2$. The wavelength in medium 2 is displaced by a factor of the wavelength in medium 1, $\lambda_2 = \lambda_1 \dfrac{n_1}{n_2}$, where n_1 and n_2 are the refractive indices of medium 1 and 2, respectively.

Scattering Correction

Although the polymer is chosen so that it does not interfere in the wavelengths of interest for the indicator dye, the polymer may still scatter light significantly. There are rigorous solutions for scattering available, such as the Rayleigh approximation for small non-absorbing scatterers. The scatterer dimensions must be a twentieth of the wavelength and the ratio of the refractive indicies of the scatterers not much larger than unity. To simplify this model, the terms that are constant over wavelength are grouped, along with the refractive index terms. In actuality, the refractive index is a function of wavelength, but since the polymers chosen here are non-absorbing, and the swelling of the matrix keeps the refractive index ratio about one, this approximation works well. The constants can be grouped for a simple semi-empirical model, $\tau(\lambda) = k\lambda^{-g}$, where k and g are estimated parameters [9]. For Rayleigh scatterers, g would be 4.

This correction is used along with the background correction in an optimization routine developed in house to calculate the dye concentration. Depending on the indicator dye, the effects of the concentration of the dye, degree of crosslinking of the matrix, ionic strength, etc. carry a different weight. For phenol red, at higher dye concentrations there are two counteracting effects, the pressure generated by electrostatic repulsion and swelling of the membrane. The pressure is relieved partly by the swelling of the membrane and by a deviation in the dissociation equilibrium. These effects are modeled in the calibration of the membranes, discussed in the next section.

Calibration

Along with the concentration of the indicator dye, ionic strength of the solution, and temperature, the degree of crosslinking affects the dissociation equilibrium. This can be represented by the swelling ratio, $Q = l_w/l_d$, where l_w and l_d are the thickness of the swollen and dry films, respectively. It has been shown that the dye does not degrade when immobilized, thus allowing solution data to be used in the calibration of the membranes [8]. The swelling ratio can now be calculated from the extinction coefficients from solution-equivalent concentrations. The swelling ratios for three films are shown in Table I [8].

Table I. Average Swelling Ratios of Phenol Red Films

Phenol Red Film	Average Q	Standard Error
No. 1	1.22	0.07
No. 2	1.28	0.05
No. 3	1.43	0.07

It would be impractical to calibrate every sensor, so the parameters are estimated for one membrane, and the pK_{ao} is used as a reference for other sensors. A calibration equation can then be set up as, $pH = b_o + b_1 pK_{a(3)} + b_2 C + b_3 Q + b_4 \log([X_2]/[X_1])$. The regressed parameters are shown below in Table II [8].

Table II. Calibration Parameters for the Phenol Red Films

Parameters	Value	Standard Error
b_o	34.491	
b_1	0.996	0.051
b_2	14644.98	1061.24
b_3	1.319	0.021
b_4	-37.850	2.745

For an arbitrary membrane, the parameters b_o, b_2, and b_3 can be lumped together. This calibration yields pH values within an expectable error.

Precision and Accuracy

Recall the interpretation model with the corrections to the equilibrium constant [8],

$$pH = pk_{ao} + P \log(Cond) + M\left(\frac{1}{T}\right) + \log\left(\frac{X_2}{X_1}\right) \tag{1}$$

An error propagation analysis on equation 1 yields the precision in the pH measurements. The generic form is as follows,

$$\text{var}(pH) = \sum_{k=1}^{s} \left(\frac{\partial f}{\partial Z_k} \right)^2 \text{var}(Z_k) + \frac{1}{2} \sum_{k=1}^{s} \sum_{j=1}^{s} \text{Cov}(Z_k, Z_j) \tag{5}$$

$$k \neq j \, j \neq k$$

where f is the function, Z is a measurement parameter, and s is the total number of parameters. The covariance of the two random variables is defined as the expected value of the product of parameter k minus its mean times parameter j minus its mean. The partial derivatives with respect to conductivity and temperature are,

$\left(\dfrac{\partial f}{\partial Cond} \right)^2 = \left(\dfrac{P}{Cond} \right)^2$ and $\left(\dfrac{\partial f}{\partial T} \right)^2 = \left(\dfrac{M}{T^2} \right)^2$. The derivatives with respect to the

protonated and dissociated forms of the indicator dye are, $\left(\dfrac{\partial f}{\partial [X_2]} \right)^2 = \left(\dfrac{1}{[X_2]} \right)^2$ and

$\left(\dfrac{\partial f}{\partial [X_1]} \right)^2 = \left(\dfrac{-1}{[X_1]} \right)^2$. The concentrations of the protonated and dissociated forms

of the indicator are correlated, represented as, $E\left(([X_1] - [\overline{X}_1])([X_2] - [\overline{X}_2]) \right)$. The covariance is put in terms of absorption using the Beer-Lambert law, since this is the measurable quantity. Putting these terms into equation 5, the following equation is obtained for the variance in the pH measurement [8],

$$\text{var}(pH) = \left(\frac{P}{Cond} \right)^2 \text{var}(Cond) + \left(\frac{M^2}{T^4} \right) \text{var}(T) + \frac{\text{var}(A_{[x_2]})}{A_{[x_2]}^2} + \frac{\text{var}(A_{[x_1]})}{A_{[x_1]}^2}$$
$$+ \frac{\left(A_{[x_1]} - \overline{A}_{[x_1]} \right)\left(A_{[x_2]} - \overline{A}_{[x_2]} \right)}{\varepsilon_{[x_1]} \varepsilon_{[x_2]} \ell^2}$$

This equation accounts for the error contributions from conductivity, temperature, and the spectra of the indicator dye. The typical error in temperature measurements is within $0.1C°$ of the measured value, making the contribution from this term insignificant. The error in the spectra is controlled by the dye concentration. With a strong signal this term is also negligible. The term that may be significant is the conductivity. Typically, conductivity measurements are within 0.5%, so at low conductivity values the contribution from this term could be significant. For this reason it is important to account for the conductivity properly. If the variance in the measurement is proportional to the measurement value, then the error term will be insignificant. The precision, or consistency of an estimation, obtainable with this system is $\pm$ 0.001 pH units [8].

The accuracy of the measurement is directly proportional primarily to the errors in the estimation of the equilibrium constant. This is dependent on the calibration. The more accurate the buffers used, the more accurate the measurement. The variance of the pH measurement is estimated by dividing the residual sum of squares of the measurement data, RSSQ, by the degrees of freedom, dfr, $\text{var}(pH) = RSSQ/dfr$. The variance in the estimated pH using phenol red is shown in Table III for four films, calibrated with high-precision buffers [8].

Table III. Variance of pH Measurements with Immobilized Dyes

Phenol Red Film	Variance
No. 1	0.0015
No. 2	0.0019
No. 3	0.0026
No. 4	0.0026

Summary

A three-layered support matrix has been developed to support indicator dyes for sensing. The outer layer consists of a crosslinked polymer matrix that contains the indicator dyes chemically bonded within this layer. The inner layer is a silane layer that covalently bonds the suport matrix to the glass [4]. An acryloyl chloride layer is used to bridge the carbon chemistry of the outer layer and the silica chemistry of the inner layer. This unique design prevents leaching of the indicator dyes and provides thermal and mechanical stability. The support matrix can withstand autoclave temperatures of 120°C and high pressure, and will not peel off the glass even after being soaked for eight months.

This design provides a framework that allows for a variety of indicators, such as those sensitive to carbon dioxide, oxygen, heavy metals, and pH, to be incorporated in the support matrix. By immobilizing multiple indicators or using fiber bundles, it is possible to design sensors that can measure multiple analytes simultaneously, pH and ionic strength for example.

To interpret the spectra obtained from the sensor, a mathematical model has been developed which accounts for pertinent effects on the pH measurement due to the immobilization of the indicator dye and the sensor design [8]. These effects can be broken up into two categories, physical-chemical effects and optical effects. The physical-chemical effects of interest are temperature, ionic strength, and buffering capacity of the polymer matrix-indicator dye system. The optical corrections that need to be made are for the difference in refractive index between the solution and the membrane and scattering. The temperature and ionic strength of the solution are reflected in the pK_a of the indicator dye. It has been shown that at low ionic strength the pK_a is no longer constant [8]. At low ionic strength, the charged groups tend to neutralize themselves with the hydrogen ions, causing a buffering effect. As the concentration of the counter ions increases, the negative charges in the membrane are shielded causing less of an effect on the pK_a. The buffering can also be reduced by the swelling of the membrane, which reduces the charge and osmotic pressure in the matrix. A kinetic model, reduced to the form shown in equation 4, adequately represents these two effects [8]. The concentrations of the protonated and dissociated forms of the indicator dye are represented spectrophotometrically by the Beer-Lambert law. Due to the membrane properties, the spectra of the indicator dye are subject to shifts. As the refractive index ratio of the medium and the membrane deviates from one, a wavelength correction is needed. The wavelength can be corrected relative to a reference. That is, the wavelength in medium 2 is displaced by a factor relative to the wavelength in medium 1. Besides refractive index differences, the spectra can be shifted due to scattering from the polymer matrix. Although the polymers used are chosen to be transparent in the range of interest, they can still scatter light significantly. A semi-empirical model has been proposed which adequately corrects for these effects [8]. With this system, precessions of ± 0.001 pH units are attainable.

A calibration procedure for a given sensor has been outlined. The needed fundamental value is the pK_{ao} for a given indicator dye. Along with this value, the calibration parameters for a specific sensor can be estimated using high precision buffers. Conveniently, solution data can be used for calibration [8].

An area that needs to be studied more is the response time of the sensor. Due to the polymeric nature of the sensor, the matrix swells depending on the pH of the solution, effecting the equilibrium and response times. The degree of swelling is different depending on the direction of the pH change, that is, more acidic or basic. This results in different response times. There are two ways to address this issue, one of which is to vary the degree of crosslinking. The less crosslinked the matrix, the more the polymer chains act independently. The crosslinking can only be lessened to a certain degree in order to maintain mechanical stability. With the chemistry fixed, another option is to develop a dynamic model that will account for the flux of ions into the membrane. In this way, the equilibrium pH could be calculated before actually reaching equilibrium. The response time is important for control strategies used in on-line applications.

Acknowledgment

The support of Ocean Optics Inc., Optical Sensor Inc., Radiometer Medical A/S, and the USF Center for Ocean Technology is gratefully acknowledge.

Literature Cited

1. Seitz, W.R. *Anal.Chem.* **1984**,*56*,16A-34A.
2. Luo, S.; Walt, D.R. *Anal.Chem.* **1989**, *61*, 174-177.
3. Tan, W.; Shi, Z.; Smith, S.; Birnbaum, D.; Kopelman, R. *Science*, **1992**, 778-781.
4. Huang, H. PhD Dissertation, University of South Florida, Tampa, FL, **1997**.
5. Huang, H.; García-Rubio, L. H. In *Interfacial Aspects of Multicomponent Polymer Materials.* Loshe, D. J.; Russell, T. P.; Sperling, L. H., Eds. Plenum Press: New York, NY, **1997**.
6. Elmer, T. H. In *Silyated Surfaces*, **1980**, 1-30, Gordon and Breach, Science Publishers, Inc.: Newark, NJ.
7. Plueddemann, E. P. In *Silyated Surfaces*, **1980**, 31-53, Gordon and Breach, Science Publishers, Inc.: Newark, NJ.
8. Chang, H. Silvia Hsiao-Yun. Ph. D. Dissertation, University of South Florida, Tampa, FL, **1996**.
9. Kerker, Milton. *The Scatering of Light, and Other Electromagnetic Radiation*; Academic Press: New York, NY, 1969.

Chapter 15

Fluorescent Optical Fibers for Data Transmission

Hans Poisel[1], Karl F. Klein[2], and Vladimir M. Levin[3]

[1]FHN, University of Applied Sciences, Wassertorstrasse 10, 90489
Nuernberg, Germany
[2]FH–GF, University of Applied Sciences, W. Leuschnerstrasse 16, 61169
Friedberg, Germany
[3]RPC, Moskovskoe SH, 157, Tver 170032, Russia

Fluorescent optical fibers offer new possibilities for building optical bus systems or rotary joints. They allow for a lateral coupling of information to a dye doped polymer fiber. Selecting the right parameters concerning lateral launch, absorption and emission wavelengths, data transfer rates beyond 500 Mbits/s are feasible.

Introduction

Usually fluorescent polymer optical fibers and data transmission belong to different worlds; until recently fluorescent fibers have been used for sensor applications or for decorative purpose whereas data transmission is usually done with clear undoped fibers. In the following it will be demonstrated that a combination of both worlds opens the door to new applications in data transmission systems.

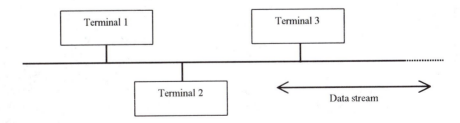

Figure 1. Data bus system: Linear topology

Background and Motivation

In data transmission systems there are a lot of applications where lateral coupling to optical fiber is needed. Optical bus systems need a coupler or tap for each terminal and this can only be realized at discrete positions.

These terminals could be lined up more easily and the system design were more flexible if lateral access were possible at any position along the fiber. It should be possible to install a terminal and to remove it without the need to install and to remove a coupler or tap. The fiber between the access points should behave as ideally as possible, i.e. no additional loss.

Another application where lateral access is desirable is an optical rotary joint or slipring. This device allows for transmission of data from a rotating device such as radar antennas, a computer tomography apparatus or from robot arms to a stationary device, e.g. a transmitter or a computer. The first description of fiber optical rotary joint with lateral access was published in 1994 (1).

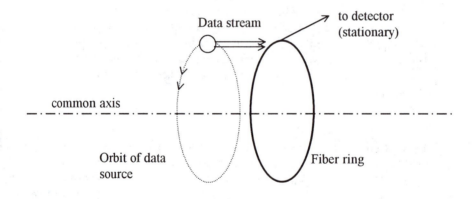

Figure 2. *Rotary optical joint*

Lateral Coupling to Optical Fibers

Apart from solutions using coupling prisms or surface gratings which are suitable preferably in the laboratory, ordinary fibers do not allow for lateral access, at least not with a considerable coupling coefficient (2). Measuring the radiation at the end of a laterally illuminated fiber yielded the results shown in Fig. 3. In addition this light is coupled mainly to high order helically or skew modes with attenuation in the range of several hundred dB/m.

A closer look to the origin of light scattering by scanning the fiber cross section with a narrow laser beam showed, that the main contribution is due to scattering at imperfections from the core – cladding interface (c.f. Fig. 4).

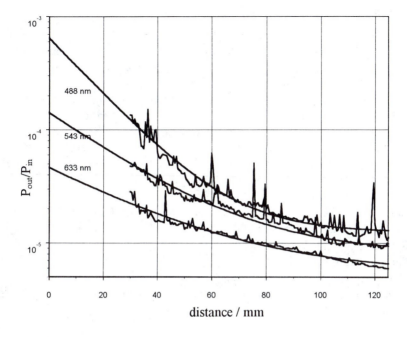

Figure 3: Ratio of radiation coupled in to incident radiation as a function of the distance between fiber end and point of illumination.

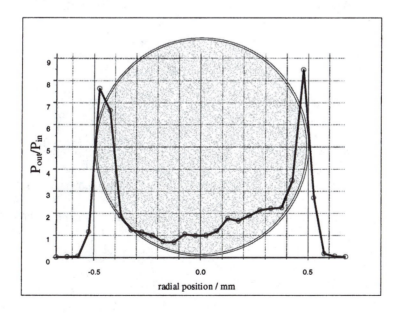

Figure 4: Power (normalized to axis value) coupled to the fiber by scanning across the fiber diameter with a spot of < 0.1 mm diameter

Therefore it was necessary to look for other solutions. One very successful solution was using dye doped polymer optical fibers, so called fluorescent optical fibers (FOF). Thus it is possible to circumvent the problem to couple **light** radially to the fiber by fulfilling the original task which is to couple the **information** to the fiber. This can be done by exciting fluorescent radiation inside the FOF by an external light source which can be modulated with the information to be transmitted. Since now the fluorescent light is generated inside of the waveguide, part of it (i.e. that part which is emitted inside the acceptance cone of the fiber giving the piping efficiency) can be trapped and guided along the fiber until to its ends.

The capabilities depend on parameters such as

- Coupling efficiency from the external light source to the FOF
- Absorption of exciting radiation
- Fluorescence yield
- Piping efficiency
- FOF attenuation for fluorescent light
- Bandwidth of the system determined essentially by fluorescent lifetime of the dopant

For the first prove of principle standard FOFs made of polystyrene had been used and data rates up to 100 Mbits/s (1) had been achieved. Based on experiences with that first demonstrator a development program for optimizing the FOF could be started.

Boundary Conditions

Because of optical and mechanical properties are the best, PMMA was chosen as a substrate material. The dyes were selected according to following criteria:

- Spectral region such that there are high bandwidth and powerful light sources (LEDs or laser diodes) available or to be expected in the near future.
- Spectral region such that the spectral attenuation of the host material is sufficiently low.
- Short fluorescence lifetime. In this case these values were only published for solution in organic solvents such as ethanol, values which may not hold for solution in polymers.
- Good solubility in selected host polymers
- Stable under normal environment e.g. daylight irradiation
- Non-toxic
- Thermally stable up to fiber extrusion temperatures

Experimental

Fibers with different dyes, concentrations and processing during extrusion have been fabricated and tested. For measuring absorption and emission spectra a set up shown in Fig. 5 was used:

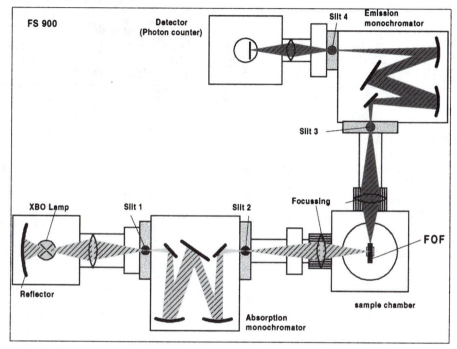

Figure 5. Set up for measuring absorption and emission spectra of fluorescent optical fibers

A 450 W XBO lamp produces sufficient white light which is coupled side-on to the FOF under test via a first monochromator controlling the absorption wavelength. The fluorescent light thus generated is analyzed by a second monochromator and detected by a thermoelectrically cooled photon counter. A PC controls the measurement sequence, processes and stores the data.

Fluorescence lifetime is measured by a laser spectrometer shown in Fig.6. A nitrogen laser pumps a dye laser producing short pulses of about 0.3 ns half width. These pulses are focussed to the FOF under test with a repetition rate of about 10/s. The fluorescence light is captured optically and coupled to a monochromator where the spectral regime to be detected by a following photomultiplier tube can be tuned. The signals of this PMT are fed to a boxcar card inside a PC where the repetitive signals are averaged and analyzed. For enhancing the time resolution, the output signal is deconvoluted with the known input signal thus giving a resolution in the range better than 0.1 ns.

FL900 CDT

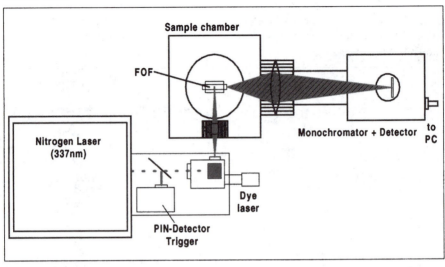

Figure 6. Set up for measuring fluorescent lifetime of dye doped optical fibers

Results and Discussion

Experimental results are shown for one of the most promising dyes: Nile blue. This dye showed a very good solubility in MMA and was stable under polymerization and extrusion conditions.

As expected, the absorption maximum is around 630 nm, the wavelength, where laser diodes as excitation sources are commercially available. Emission maximum is around 700 nm, unfortunately close to an absorption peak of PMMA. For testing the reproducibility the sample was rotated along its axis giving a variation of less than 10% due to imperfection of end face preparation of the fiber sample. Emission and absorption curves overlap in the range between 600 and 700 nm giving reasonable self-absorption of fluorescent light. This self-absorption is expected to prolong the fluorescence lifetime.

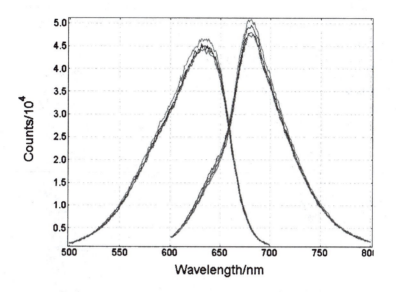

Figure 7. Absorption and emission spectrum of Nile blue doped PMMA fiber (sample AC-K). Different traces show dependence on sample orientation.

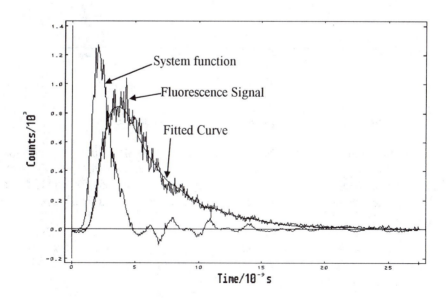

Figure 8. Excitation pulse and response function of Nile-blue doped sample K

The dynamic behavior for this sample can be seen in Fig.8. After deconvolution with the system function a fluorescence lifetime of 3.5 ± 0.2 ns was calculated which is very fast for a red emitting dye (3)

The concentration was varied by a factor of 10 but within the measurement accuracy only a slight increase of fluorescence lifetime could be detected, indicating only a small concentration dependence if any (Fig.9). To investigate the influence of self absorption, the wavelength of the output monochromator was varied, but again no influence was noted.

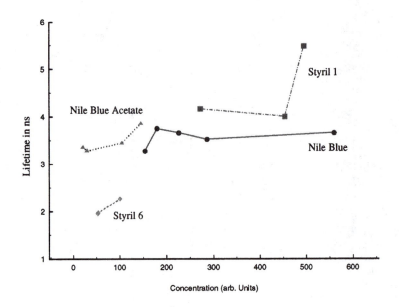

Figure 9: Fluorescence lifetime as function of dopant concentration

Several different dyes have been tested, most of them with different concentrations or different treatment during the polymerization and extrusion process. Some of the originally selected dyes did not solve in the monomer, some reacted chemically with the monomer and some were not stable under the conditions of polymerization.

Those dyes which gave a reasonable fluorescence signal are listed in Table I.

The shortest lifetime measured with a good signal amplitude was 1.9 ± 0.2 ns. This allows data transmission bandwidth in the order of 500 Mbits/s and beyond, depending on the modulation scheme.

Table I. Fluorescent Lifetimes of selected Dyes*

Dye	λ Emission max. (nm)	λ Absorption max (nm)	Lifetime (ns)
Nile blue	633	672	3.5
Oxazine 1	646	670	4.2
Styril 9M**	585	840	1.3
Styril 6	615	720	1.9
LD 700	643	700	3.8
LD 688	492	610	2.8
Oxazine 750**	667	750	2.9

* all dyes from Lambdachrome, Goettingen, Germany
** very weak signal

Conclusion

Fluorescent optical fibers have been investigated for data transmission applications because of their very attractive property, to couple information laterally to the waveguide. Different fluorescent dopants have been tested showing that data transmission systems with data rates beyond 500 Mbit/s are feasible. More investigations have to be done to enhance the concentration of the most promising dyes in order to achieve data rates in the Gbit/s regime.

Acknowledgment

Parts of this work have been supported by the German Ministry BMB+F contract # 1704897. The authors would like to thank Oliver Stefani and Markus Beck for most valuable contributions.

References

(1) Poisel, H., Dandin, E., Klein, K.F., *Proceedings of 4th International Conference on Polymer Optical Fibers POF'94, Yokohama, Japan,* **1994**, *82.*
(2) Poisel, H., Hager, A., Levin, V., Klein, K.F., *Proceedings of 7th International Conference on Polymer Optical Fibers POF'98, Berlin, Germany,* **1998**, *114.*
(3) P.A. Cahill, *Radiat. Phys. Chem. Vol.41,* **1993,** *pp. 351 – 363*

Chapter 16

Polymer Scintillators: Continuous versus Intermittent Gamma Irradiation Effects

E. Biagtan[1], E. Goldberg[2], R. Stephens[3], E. Valeroso[4], M. Calves[5], and J. Harmon[5,*]

[1]Guidant Company, 26531 Ynez Road, Temecula, CA 92591
[2]Department of Materials Science and Engineering, University of Florida, MAE 314, Gainesville, FL 32611–2085
[3]Department of Physics, University of Texas at Arlington, Arlington, TX 76019
[4]Calloway Company, 2285 Rutherford Road, Carlsbad, CA 92009
[5]Chemistry Department, University of South Florida, 4202 East Fowler Avenue, Tampa, FL 33620–5250

This study focuses on phenyl containing optical polymers doped with fluorescent dyes for use as scintillator materials. The purpose is to test the effect of intermittent versus continuous doses of gamma irradiation on the light transmission and harvesting capacities of polymer scintillators. Three polymer scintillators were irradiated to a total dose of 10 Mrad. Selected samples were irradiated continuously to the total dose, while other samples were allowed to recover after every 2 Mrad. Compared to continuous irradiation, intermittent irradiation produced less transient light output losses, but produced the same amount of permanent losses. The differences in degradation behavior between the continuously and intermittently irradiated samples varied with the dose and dose rate.

Background

Polymeric scintillators are used in detectors for particle accelerators, radiobiology, and astrophysics to monitor ionizing radiation. Polymeric scintillators are used to monitor ionizing radiation in detectors used in particle accelerators, radiobiology and astrophysics (1). Recently, relatively good results have been achieved in low-energy x-ray imaging experiments (2). Optical polymers offer advantages over other scintillators materials. They are easier to process into fibers and plates than inorganic scintillator and more convenient to use that liquid scintillator in many applications. Plastic scintillator is flexible and immune to electromagnetic interference. In addition, many optimum fluorescent scintillator dyes are soluble in phenyl containing polymers and these dyes exhibit rapid response times often less that 5 nanoseconds.

Scintillation in organic molecules is the process wherein radioactive particles excite π electrons in phenyl rings; photons are emitted when the excited electrons relax to their ground state. These photons travel through the scintillating material which is in plate or fiber form. They are transported through the medium to photodetectors where they are counted. The most efficient way to transport these photons is via a cascade process. That is, the photons emitted by the polymer, usually in the region of 280-320 nm, are absorbed by a primary fluorescent dye and re-emitted at a longer wavelength. A second dye absorbs photons of the energy emitted by the primary dye. These photons are emitted at a still higher wavelength. The purpose of the process is to optimize the final emitted wavelength to match that of the photon detection system and to avoid unwanted absorptions in the polymer often encountered at lower wavelengths. In some scintillator one, large Stokes Shift dye replaces the primary and secondary dye (3).

The performance and efficiency of polymeric scintillators is directly related to radiation induced reactions that occur in the polymer structure and in the dyes (4). Dye degradation can lead to destruction of the chromophore or a shift in the absorption emission maximum. The primary effect of radiation on the optical properties of glassy polymers is the formation of color centers. These color centers absorb the photons that are emitted by scintillation and thus, diminish scintillator efficiency. They are of two types, permanent color centers and transient or annealable color centers (5,6) Permanent color centers are radiation induced conjugated systems that absorb light in the UV/visible region of the electromagnetic spectrum. They are formed from the reaction of atoms in the scintillator structure. If oxygen is present in the matrix, carbonyl compounds may be incorporated into the unwanted chromophore structure. Transient color centers are radiation induced free radicals. These free radicals are colored species that remain locked into the stiff polymer matrix. In flexible systems the free radicals annihilate by recombination. (7). In stiff polymer matrices the free radicals persist until they are quenched by oxygen or undergo a chance recombination reaction. Since annealing is associated with the presence of oxygen in the polymer matrix, annealing may occur during irradiation and after irradiation and is effected by the rate in which oxygen diffuses into the matrix (8-10).

The availability of oxygen at the free radical sites is responsible for dose rate dependency that is noted when scintillator is irradiated. At lower dose rates oxygen diffuses into the matrix during irradiation and reacts with radicals that may have undergone side reactions, or recombination if no oxygen is present. Thus, the lower the dose rates the higher the amount of permanent color centers. We have shown that this is directly related to the degradation of light output in polymeric scintillators (3,11-13). In addition, scintillation detectors are often used intermittently. It seems likely that when scintillator is left unexposed to radiation in between experiments and is allowed to recover during these intervals it will have a different oxygen profile and therefore different light output losses as compared to scintillator continuously irradiated to the same dose. With this in mind, scintillators were subjected to intermittent irradiation and their changes in light output were compared to those for continuously irradiated scintillators.

Experimental Details

Three commercial polymer scintillators were used, SCSN-38, SCSN-81, and Bicron-499-35. SCSN-38 and SCSN-81 were obtained from Kurarey Company Ltd. Bicron 499-35 was obtained from Bicron Corporation. SCSN-38 is composed of optical grade polystyrene doped with 1.0% by weight 2-phenyl-5-(4-biphenylyl)-1,3,4-oxadiazole) and 0.02% by weight 1,4-bis (2,5-dimethylstyryl) benzene. SCSN-81 is composed of optical grade polystyrene doped with 1.0% by weight p-terphenyl and 0.02% 2,5-bis[5-tert-butylbenzoxazolyl (2)]thiophene. The exact composition of Bicron-499-35 is proprietary. However, the primary dye in the Bicron scintillator is p-terphenyl and the secondary dye is 2,5-bis[5-tert-butylbenzoxazolyl (2)] thiophene. The base polymer is a copolymer containing polyvinyltoluene. The scintillators were obtained as 4 mm thick plates and were machined into 1" OD disks. They were irradiated with a ^{60}Co gamma source while exposed to air. A computer program determined the required exposure durations based on the distance from the source, the desired gamma dose, and the starting date for irradiation.

The light yield of a disk was measured by placing a 1.0 μCurie Am-241 alpha source on top of the disk. The disk was then coupled with optical grease to a photomultiplier tube (THORN EMI type 9124B operating at 1100 V). Scintillations were collected for one minute and downloaded into a computer as a spectrum of channel numbers versus counts. The channel number with the highest count was taken as the light yield of the sample. The light output of the sample was the ratio of its light yield versus the light yield of an overall standard, an un-irradiated disk of SCSN-38. Their respective pre-irradiation light outputs of SCSN-38, SCSN-81 and Bicron 499-35 are 100- 105% (±5%).

In an earlier study, sets containing five samples of each scintillator had been continuously irradiated to 2, 4, 6, 8, and 10 Mrad (13). Separate sets were irradiated at different dose rates ranging from 1.5 Mrad/hr to 0.0023 Mrad/hr. Their light outputs were measured before irradiation (pre-irradiation LO), just after the desired dose, and daily thereafter until stable (within 7 days after irradiation). During this 7 day recovery period, the samples were stored in a dark non-airtight container. The amount of recovery is the difference between the light output after the recovery period (post-recovery LO) and the light output just after irradiation (post-irradiation LO). Some of the results of this earlier study are presented here for comparison. In the present study, two samples of each scintillator were irradiated to a total dose of 10 Mrads, but after every 2 Mrad they were removed from the irradiator and allowed to recover for 7 days. One sample was irradiated at 1.5 Mrad/hr and the other at 0.14 Mrad/hr. Light outputs were measured before irradiation, just after every 2 Mrad, and daily during each recovery period.

Results and Discussion

Continuous irradiation

Figs. 1a-d are plots of the post-irradiation and post-recovery LOs versus the irradiation dose for the SCSN-81 samples that were irradiated continuously(13). The dose rate of irradiation decreased going from Fig. 1a to 1d. The bars and numerical values represent the amount of recovery. There are several general trends with the dose and dose rate that can be observed. At a fixed dose rate, both the post-irradiation LO and post-recovery LO decreased with increasing dose. The former decreased more rapidly than the latter; the amount of recovery increased with increasing dose.

Note though, that at each of the stated doses, the amount of recovery decreased with decreasing dose rate of irradiation until there were no discernible differences between the post-irradiation and post-recovery LOs (fig. 1d). Earlier, via ESR experiments, we explained the dose rate effect on light output losses in terms of radical production rate, secondary reaction rates, and oxygen diffusion (12). At high dose rates, the radical production rate is greater than their conversion rate, and oxidation reactions are not sustained by diffusion, so many trapped radicals remain after irradiation. These remaining radicals are annealed, most likely by recombination, without forming additional permanent color centers. This is due to the fact that at higher dose rates ESR cause more dense population of free radicals, which is indicated by ESR signals. The proximity of one free radical to another increases the chance of recombination. Therefore, at high dose rates there is a large amount of recovery and little permanent loss in the light output. At lower dose rates, the production rate of radicals decreases, more time is available for secondary reactions, and oxidation reactions are sustained by diffusion, so fewer trapped radicals remain after irradiation. This results in less recovery and more permanent losses in light output.

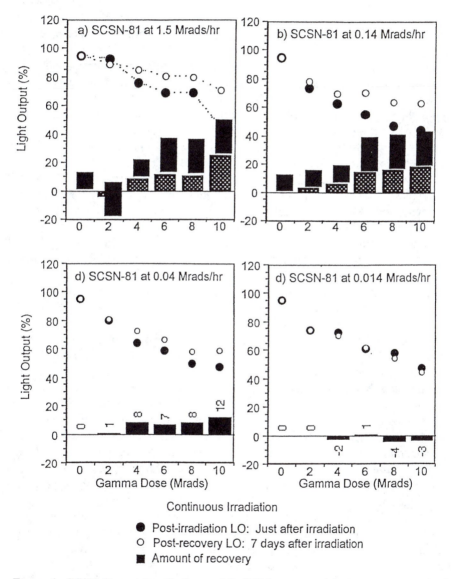

Figure 1. SCSN-81 post-irradiation and final light output with gamma irradiation to various dose rates and doses.

Intermittent Irradiation

Figs. 2, 3 and 4 are plots of the post-irradiation and post-recovery LOs versus the gamma dose for the SCSN-38, SCSN-81 and Bicron 499-35 samples that were intermittently irradiated at dose rates 1.5 Mrad/hr and 0.14 Mrad/hr. Comparisons between the intermittent and continuous irradiation data must exclude the values at 2 Mrad, since the irradiation conditions were the same up to that point. Any differences in the light outputs for the samples at 2 Mrad are attributed to data scatter.

Analysis of variance tests indicated that the effects of intermittent irradiation on the post-irradiation LO was significantly different from the effects of continuous irradiation (p-values are 0.0233 for SCSN-39, 0.0011 for SCSN-81, and 0.0198 for Bicron-499-35), p is the probability that continuous and intermittent LO values belong to the same data set. As expected, continuous and intermittent samples at both dose rates exhibited permanent light output losses that increased with increasing doses and decreased with dose rate.

At 1.5 Mrad/hr, all three scintillators had much higher post-irradiation LOs with intermittent irradiation. This is consistent with the effect of color center proximity on recombination at high dose rates. Post recovery light outputs were the same for continuous and intermittent samples at the higher dose rate. The differences between intermittent and continuous irradiation were scintillator dependent.

Bicron-499-35 scintillator studies at a dose rate of 1.5 Mrad/hr exhibited less loss in light output after irradiation in intermittent tests (fig. 4). Both intermittently and continuously irradiated samples recovered to 90% light output a 10 Mrad doses after a 1 week recovery period. Samples irradiated at 0.14 Mrad/hr displayed different behavior. That is, both continuous and intermittent samples had the same post irradiation light output. The intermittent sample had more permanent damage, recovering to 52% light output versus 65 % for the continuously irradiated sample. Permanent light output loss increased as the dose rate decreased. Samples irradiated at 1.5 Mrad/hr exhibited a permanent loss of 90% after recovery.

ESR results on Bicron-499-35 taken from ref. 12. Are shown in fig. 5. The free radical population and recovery speed increases with dose rate. This data explains the results on the Bicron samples; the ease in free radical recombination increases as the population increases. At the lower dose rate, recovery depends more heavily on oxygen quenching than on recombination. In the intermittent samples, there was more permanent damage due to the fact that more oxygen diffuses into the matrix to react, quench and form carbonyl compounds.

The SCSN-81 samples (fig. 3) showed a different dependency on continuous versus intermittent irradiation. At the higher dose rate the intermittent sample recovered during intermittent aging times. Both continuous and intermittent samples recovered to 75% light output. Unlike the Bicron samples, at the lower dose rate, the intermittent sample showed less transient color centers than the

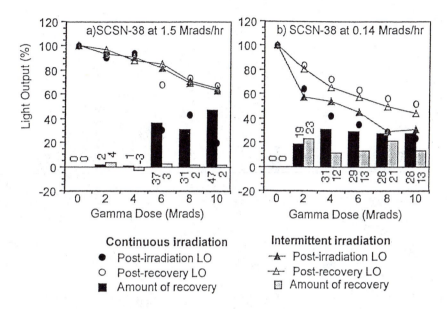

Figure 2. SCSN-38: Continuous versus intermittent irradiation.

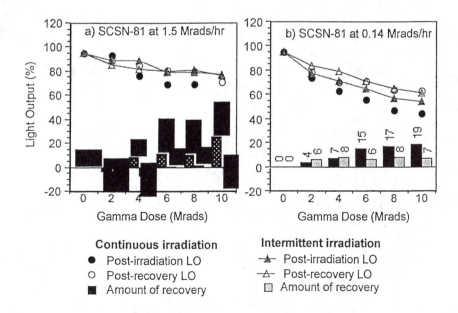

Figure 3. SCSN-81: Continuous versus intermittent gamma irradiation.

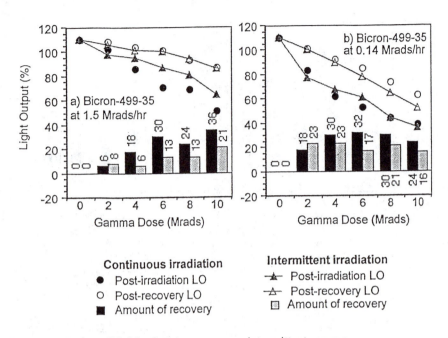

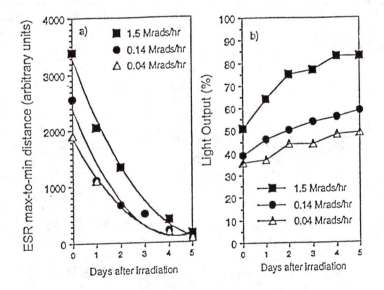

Figure 4. Bicron-499-35: Continuous versus intermittent gamma irradiation.

Figure 5. Bicron-499-35: (a) ESR intensities and (b) LO during the recovery period after gamma irradiation to 10Mrad at various dose rates.
*Reproduced with permission from reference 12. Copyright 1996.

continuously irradiated samples. However, all SCSN-81 samples recovered to only 65% light output as compared to 75% at the higher rate.

SCSN-38 samples (fig. 2) behaved similarly to SCSN-81 samples in that intermittent samples annealed during the radiation cycling. All samples irradiated at 1.5 Mrad/hr recovered to 70& light output. The samples tested at 0.14 Mrad/hr exhibited the most permanent color centers. After 10 Mrad, continuous samples have 52% light output, while intermittent samples had 47% light output.

Conclusion

Intermittent irradiation of polymer scintillators results in different post-irradiation LOs compared to continuous irradiation. The results vary with the scintillator brand and the dose rate, but they are transient. At 10 Mrad doses administered at a dose rate of 0.14 Mrad/hr, SCSN-38 and the Bicron samples exhibited more permanent color centers in intermittent tests. At the high dose rate all continuous and intermittent samples recovered to the same % light output for each type of scintillator.

All of this reveals that accelerated testing experiments on scinitllator need to be designed with great consideration for diffusion effects. Dose rate dependency arises from oxygen diffusion rates in the polymer matrix. In detector applications dose rates are low and intermittent use is also a factor. This may lead to higher permanent light losses . Again, these results are scintillator dependent.

Acknowledgements
The authors are grateful to Dr. H. Hanrahanm at the University of Florida for use of the ^{60}Co gamma irradiator.

References

1. Rebourgeard, P., Rondeaux, F., Baton, J. P., Besnard, H., Blumenfeld, H., Bourdinaud, M., Calvet, J., Cavan, J., Chipaux, R., Giganon, A., Heitzmann, J., Jeanney, C., Micolon. P., Neveu, M., Pedrol, T., Pierrepont, D., and Thevenin, J. *Nucl. Intrum. And Methods in Phys. Res.* **1999**, *A427*, 9. 543.
2. Ikhlef, A. and Skowronek, M. *Apply. Optics,* **1998**, *37 No. 4*, p. 8081.
3. Biagtan, E., Goldberg, E., Stephens, R., Valeroso, E., and Harmon, J. *Nucl. Intrum. And Methods in Phys. Res.* **1996**, *B114*, p. 88.
4. Busjan, W., Wick, K., and Zoufal, T. *Nucl. Intrum. And Methods in Phys. Res.* **1999**, *B152*, p. 89.
5. Wallace, J., Sinclair, M., Gillen, K., and Clough, R., *Raiat. Phys. Chem.***1993**, *41 No. ½*, p. 85.

6. Gillen, K. T., and Wallace, J., and Clough, R., *Raiat. Phys. Chem.,***1993**, *41 No. ½*, p. 101.

7. Harmon, J. P. and Gaynor, J., *J. Polym. Sci. Part B: Polym. Phys.***1993**, *31*, p. 235.

8. Gillem K. and Clough, R., *J. Polym. Sci. Polym. Chem. Edu.***1985,** *23*, p. 2683.

9. Gillen K. and Clough, R., *Polym. Deg. And Stab.,* **1989**, *24*, p. 137.

10. Harmon, J., Biagtan, E., Schueneman, G. and Goldberg, E., in *Irradiation of Polymers, ACS Symposium Series 610.* Eds., R. Clough and S. Shalaby, Washington DC, **1994,** p. 302.

11. Biagtan, E., Goldberg, E., Harmon, J. and Stephens, R., *Nucl. Intrum. And Methods in Phys. Res.* **1994**, *B93*, p. 296.

12. Biagtan, E., Goldberg, E., Stephens, R. and Harmon, J. Nucl. Intrum. And Methods in Phys. Res. **1996**, B114, p. 302.

13. Biagtan, E., Goldberg, E., Stephens, R., Valeroso, E. and Harmon, J. Nucl. Intrum. And Methods in Phys. Res.**1996**, B114, p. 125.

Author Index

Subject Index

Highlights from ACS Books

Desk Reference of Functional Polymers: Syntheses and Applications
Reza Arshady, Editor
832 pages, clothbound, ISBN 0–8412–3469–8

Chemical Engineering for Chemists
Richard G. Griskey
352 pages, clothbound, ISBN 0–8412–2215–0

Controlled Drug Delivery: Challenges and Strategies
Kinam Park, Editor
720 pages, clothbound, ISBN 0–8412–3470–1

A Practical Guide to Combinatorial Chemistry
Anthony W. Czarnik and Sheila H. DeWitt
462 pages, clothbound, ISBN 0–8412–3485–X

Chiral Separations: Applications and Technology
Satinder Ahuja, Editor
368 pages, clothbound, ISBN 0–8412–3407–8

Molecular Diversity and Combinatorial Chemistry: Libraries and Drug Discovery
Irwin M. Chaiken and Kim D. Janda, Editors
336 pages, clothbound, ISBN 0–8412–3450–7

A Lifetime of Synergy with Theory and Experiment
Andrew Streitwieser, Jr.
320 pages, clothbound, ISBN 0–8412–1836–6

For further information contact:
Order Department
Oxford University Press
2001 Evans Road
Cary, NC 27513
Phone: 1-800-445-9714 or 919-677-0977
Fax: 919-677-1303

Bestsellers from ACS Books

The ACS Style Guide: A Manual for Authors and Editors (2nd Edition)
Edited by Janet S. Dodd
470 pp; clothbound ISBN 0–8412–3461–2; paperback ISBN 0–8412–3462–0

Writing the Laboratory Notebook
By Howard M. Kanare
145 pp; clothbound ISBN 0–8412–0906–5; paperback ISBN 0–8412–0933–2

Career Transitions for Chemists
By Dorothy P. Rodmann, Donald D. Bly, Frederick H. Owens, and Anne-Claire Anderson
240 pp; clothbound ISBN 0–8412–3052–8; paperback ISBN 0–8412–3038–2

Chemical Activities (student and teacher editions)
By Christie L. Borgford and Lee R. Summerlin
330 pp; spiralbound ISBN 0–8412–1417–4; teacher edition, ISBN 0–8412–1416–6

Chemical Demonstrations: A Sourcebook for Teachers, Volumes 1 and 2, Second Edition
Volume 1 by Lee R. Summerlin and James L. Ealy, Jr.
198 pp; spiralbound ISBN 0–8412–1481–6
Volume 2 by Lee R. Summerlin, Christie L. Borgford, and Julie B. Ealy
234 pp; spiralbound ISBN 0–8412–1535–9

The Internet: A Guide for Chemists
Edited by Steven M. Bachrach
360 pp; clothbound ISBN 0–8412–3223–7; paperback ISBN 0–8412–3224–5

Laboratory Waste Management: A Guidebook
ACS Task Force on Laboratory Waste Management
250 pp; clothbound ISBN 0–8412–2735–7; paperback ISBN 0–8412–2849–3

Good Laboratory Practice Standards: Applications for Field and Laboratory Studies
Edited by Willa Y. Garner, Maureen S. Barge, and James P. Ussary
571 pp; clothbound ISBN 0–8412–2192–8

For further information contact:
Order Department
Oxford University Press
2001 Evans Road
Cary, NC 27513
Phone: 1-800-445-9714 or 919-677-0977